EXERCISES IN MODERN MATHEMATICS

D. T. E. MARJORAM

PERGAMON PRESS

OXFORD · LONDON · EDINBURGH · NEW YORK
TORONTO · SYDNEY · PARIS · BRAUNSCHWEIG

Pergamon Press Ltd., Headington Hill Hall, Oxford
4 & 5 Fitzroy Square, London W.1
Pergamon Press (Scotland) Ltd., 2 & 3 Teviot Place, Edinburgh 1
Pergamon Press Inc., 44–01 21st Street, Long Island City, New York 11101
Pergamon of Canada Ltd., 6 Adelaide Street East, Toronto, Ontario
Pergamon Press (Aust.) Pty. Ltd., 20–22 Margaret Street, Sydney, New South Wales
Pergamon Press S.A.R.L., 24 rue des Écoles, Paris 5e
Vieweg & Sohn GmbH, Burgplatz 1, Braunschweig

First edition 1964 Reprinted 1965
Reprinted with corrections 1966

Printed in Great Britain by Blackie & Son Ltd., Glasgow

(2067/64)

Marian Davies.

MD

EXERCISES IN MODERN MATHEMATICS

ACKNOWLEDGEMENTS

MUCH of my own enthusiasm has been fired by the imaginative efforts of gifted mathematicians and schoolmasters working in these new and uncharted fields.

Most of the exercises in this book have been used in the classroom at Doncaster. Without the whole-hearted co-operation and encouragement of my own headmaster and mathematics colleagues, the writing of this book would have been almost impossible. In particular I wish to acknowledge the continued encouragement and valuable advice of a member of H.M. Inspectorate.

I am obliged to H.M.S.O. for permission to quote certain data from the *Monthly Digest of Statistics*. Finally, I wish to express my thanks to the Pergamon Press, to Mr. W. J. Langford for his helpful comments, to the editors who have assisted me during the various stages of preparation, and to the compositors for the excellence of their work.

The third impression of this book incorporates a number of minor corrections, additions and alterations, some of which have been made following helpful suggestions gratefully received from readers of earlier impressions of the book.

June 1966 D.T.E.M

CONTENTS

INTRODUCTION ix

1. SETS 1

Describing and listing a set; elements of a set; the empty set; equal sets; subsets; the intersection of two sets; the union of two sets; the complement of a set; Venn diagrams; the laws of sets; problems

2. SETS OF POINTS 18

Ordered pairs; lines and regions; relations; functions; mappings; successive mappings; the domain and range of a function

3. LINEAR PROGRAMMING 38

Miscellaneous problems including the transport problem

4. SETS, LOGIC AND SWITCHING CIRCUITS 47

Premises, conclusions and Venn diagrams; sentence logic; truth tables; Boolean algebra; application to logic; switching circuits

5. SCALES OF NOTATION 77

The binary scale; converting denary numbers to binary numbers; addition and subtraction of binary numbers; multiplication and division of binary numbers; binary fractions; use of the binary scale; the binary slide rule; other scales of notation

6. GROUPS 96

Symmetry and the alphabet; groups in geometry, the group of the rectangle; groups in arithmetic; groups in algebra; subgroups; isomorphic groups

7. MATRICES 115

A matrix as an operator; multiplication of matrices; matrix multiplication in arithmetic; triple matrix products in geometry; equality of matrices; addition of matrices; the inverse matrix

8. VECTORS 142

Column vectors and row vectors; vector addition; free vectors; applications in elementary geometry; more applications of the vector law of addition—vector quantities; the scalar product of two vectors; further applications to elementary geometry; unit vectors and co-ordinate geometry

9. PROBABILITY AND STATISTICS 172

Collection of data; representation of data—pie charts and bar charts, frequency polygons—histograms; analysis of the data—mean, median and modal values; dispersion or "spread"; the normal distribution curve; probability and chance.

10. TOPOLOGY 194

Oranges and doughnuts; twisted surfaces; faces, edges and corners; networks; coloured bricks and patchwork quilts

ANSWERS 217

INTRODUCTION

SCHOOL mathematics today is passing through an extremely interesting stage of development. Syllabus content and methods of teaching at all levels are undergoing a widespread and searching re-examination. So called "modern" mathematical topics, which until recent years were taught only in the universities, are now being introduced into many secondary school classrooms and school syllabuses.

Although this movement towards reform is international, it probably dates, in our own country, from the Southampton Conference of 1961 under the chairmanship of Professor Bryan Thwaites. Since that conference, able and imaginative teachers of mathematics in different parts of the country have been conducting a variety of interesting classroom experiments with revised syllabuses. In some cases groups of schools are working together from experimental texts, and several examination boards have co-operated and agreed to set O-level examinations on the revised syllabuses. Notable among these "group" experiments are the Southampton Mathematics Project, the Midlands Mathematical Experiment and the Leicestershire Schools Psychology and Mathematics Project. In other cases individual schools have tackled the problem independently, and among these the work which has been done at St. Dunstan's College, Catford, is now well known.

Most of the work done so far has been concerned with G.C.E. examination syllabuses, but experiments with mathematical curricula for other pupils have been carried out, and seem likely to increase with the advent of C.S.E. examinations and the formation of local and regional subject panels of teachers.

The movement towards syllabus reform in schools has been encouraged and assisted by the universities, the training colleges and, of course, the Ministry of Education. Courses and conferences on the "new mathematics" for teachers of mathematics in both secondary and primary schools tend to be over-subscribed, and have increased rapidly in number and popularity. Published works on modern mathematical topics and the ways in which these can be presented at an elementary level remain limited. Two of the major projects have published reports and draft chapters of lesson notes and exercises, and, of course, the Mathematical Association and the Association of Teachers of Mathematics have published a considerable number of illuminating articles in their respective journals.† Further, the Ministry of Education has circulated a full and informative report on the Cambridge Conference, held in July 1963, to all secondary schools. Nevertheless, there are few textbooks available, and the teacher who wishes to introduce such topics as the algebra of sets and its applications, simple groups, matrices, vectors, linear programming, binary arithmetic and elementary topology into his classroom, is faced with the task of devising his own supplementary lesson notes, exercises and solutions. Furthermore, such new-course books as are available come in the form of a "package deal". They include

† Since this introduction was written the following "modern" mathematical classroom texts have been published:

- *The Midlands Mathematical Experiment*, Book 1, Harrap.
- *Understanding Mathematics*, Book 1, by R. R. Skemp, U.L.P.
- *The School Mathematics Project*, Book T, C.U.P.
- *The St. Dunstan's College Booklets*, ed. G. Matthews, Edward Arnold.
- *Modern Mathematics for Schools*, Scottish Mathematics Group, Blackie and Chambers.
- *Learning Mathematics* (The Shropshire Mathematics Experiment) R. S. Heritage—Penguin.

New books for teachers include:

- *Some Lessons in Mathematics*, ed. T. J. Fletcher, C.U.P.
- *Fundamental Concepts of Mathematics*, by R. L. Goodstein, Pergamon.
- *The Core of Mathematics*, by A. J. Moakes, Macmillan.
- *Modern Mathematics in Secondary Schools*, by D. T. E. Marjoram, Pergamon.

new topics among others which are already covered by standard texts. To be fair, it should be admitted that, at its best, the spirit of the new mathematics lies as much in its new and unified approach to familiar topics as in its inclusion of new topics. Even so, many teachers of mathematics may be loth to dispose of large and expensive stocks of perfectly sound but traditional mathematical texts. Others, while not averse to experiment, may prefer to introduce new topics in their own time and manner.

It was with this purpose in mind that *Exercises in Modern Mathematics* was written. The book includes all the new topics which have appeared, or are likely to appear, on the new G.C.E. O-level and C.S.E. syllabuses. The great majority of the exercises have been used with classes in the Doncaster Technical Grammar School for Boys and, where necessary, revised in the light of that experience.

It is not intended that a pupil should work solidly through this book, or, indeed, that any one chapter should be covered completely before proceeding to other work. Some teachers, for example, may wish to use the first exercise or two in each chapter during the first year, later exercises in subsequent years and the last exercise in the fourth or fifth years. Others may prefer to take some topics in the first year and to leave others for later years. The most difficult exercises in each chapter are approximately of G.C.E. O-level standard, although in Doncaster we have used the more difficult exercises with sixth-form pupils on a course designed to supplement their normal A-level work.

So great is the variety of possible syllabuses, children's abilities and teachers' preferences, that any indication of the manner in which the book should be used has been deliberately avoided. Its main object is to provide a large selection of exercises and answers on the new topics, thereby saving teachers some of the labour of constructing their own. It will perhaps encourage others to "have a go", while allowing to all the widest freedom and discretion in its use.

At the same time it will be noted that there is a certain unity about the book and a purpose behind the order of chapters. Starting with the most elementary concept of all—that of a set, we define the operations of taking intersections, unions and complements, and develop the algebra which results from these definitions. From this we proceed to the Cartesian product and sets of ordered pairs. This leads on naturally to lattices, subsets of the Cartesian plane, lines and equations, regions and inequalities, relations, functions and mappings, and finally, in Chapter 3, to the solution of simple problems in linear programming.

In Chapter 4 we show that the algebras of statements and of switching circuits are isomorphic with the algebra of sets and are examples of Boolean algebra. Experience has shown that practical applications of this sort add point to the ideas of Chapter 1 and appeal to pupils who tend to learn better through the applications of the subject. Moreover, Boolean algebra is quickly mastered, and boys, particularly, enjoy designing specified switch circuits via the required closure table and Boolean function.

Chapter 5 is concerned with the use of binary arithmetic in computing and introduces other scales of notation. The idea of a group is introduced geometrically through the movements of a rectangle, and later, examples of modulo arithmetic and groups of functions are given. Essentially a group is a special kind of set where a law of combination defined over the elements gives rise to certain special properties.

Chapter 7, on matrices, is mainly concerned with the very special group of 2×2 non-singular matrices and their function as operators in the Cartesian plane.

Chapter 8 develops vector algebra from the work of Chapter 7 and stresses the equivalence between matrix and vector addition, matrix multiplication and the scalar product of vectors.

Chapter 9 is concerned with the arithmetical analysis of sets of data and the elementary ideas of probability.

It might be thought that Chapter 10, on topology, is hardly a matter for serious classroom study, and many of the exercises are of a qualitative rather than a quantitative nature. The fact remains, however, that of all the "modern" mathematical topics now being taught in schools, this one is truly modern. It is that branch of pure mathematics which has developed most promisingly and most rapidly in the last twenty years. Point set topology has revolutionized the pure mathematician's approach to analysis, and many of the "classical" problems of topology—the Königsberg bridges, the five-colour map and the behaviour of Möbius strips—make fascinating puzzles for the young. It is right that children should be aware that such problems are capable of mathematical treatment, and so it is hoped that even the most cautious of teachers will risk a few periods on the subject, albeit at the end of term or after the summer examinations.

Lessons on work of this nature involve a good deal of discussion between the teacher and his class. In fact many will find that the new topics generally require a good deal of discussion. Although the concepts involved are simple and delightfully interconnected, the vocabulary and notation may present initial difficulties. One of the best ways to overcome these is to encourage pupils to use the new terms freely. The child who can talk about ordered pairs, mappings and isomorphism, is well on the way to understanding what they mean.

The exercises throughout the book are linked by explanation and discussion. It is hoped that this has been done in such a way that the most intelligent children will be able to learn from the book and proceed at their own pace.

1
SETS

DESCRIBING AND LISTING A SET

A set is a collection of objects of the same type. You will have used this word already to describe a collection of things used for a common purpose, e.g. a set of spanners, a set of table mats, a set of mathematical instruments, a set of stamps and so on. In mathematics we extend the meaning of the word "set" to include cases such as the following:

The desks in your classroom form a set.
The children in your school form a set.
The pitches on your playing field form a set.
The weights for a laboratory balance form a set.
The digits 1, 2, 3, 4, 5, 6, 7, 8, 9, 0 form a set.
The questions in the first exercise in this book form a set.

A set may include so many objects that it is impossible to write them all down even if one had unlimited time.

Examples. (i) The set of all the natural numbers 1, 2, 3, 4 . . .
(ii) The set of all the fractions between 0 and 1.

Instead of *describing* what sort of objects a set contains we can make an actual list of them. We write this list in curly brackets. { } means "the set of", e.g. the set of all the days in the week whose names begin with *T* is {Tuesday, Thursday}.

Exercise 1(*a*)

List the following sets:

1. The set of all the days in the week whose names begin with *S*.

2. The set of all the days on which you have mathematics lessons.

3. The set of all the months in the year whose names begin with *J*.

4. The set of all the months in the year which have 30 days.

5. The set of all the Ridings of Yorkshire.

6. The set of all the numbers which can be thrown with a die (or dice).

7. The set of all the planets nearer to our Sun than the Earth.

8. The set of all the dominoes which have a total value less than five.

9. The set of all the natural numbers between 5 and 10.

10. The set of vowels in the word "mathematics".

11. The set of even numbers between 10 and 20.

12. The set of numbers between 10 and 20 which are divisible by 3.

13. The set of even numbers between 10 and 20 which are divisible by 3.

14. The set of numbers between 10 and 20 which are either even or divisible by 3.

15. The set of numbers between 10 and 20 which are even but not divisible by 3.

Give a description of each of the following sets and supply *one* more member of each set:

16. {A, E, I, O, }.

17. {North, South, East, }.

18. {Road, street, avenue, }.

19. {1956, 1960, 1964, }.

20. {Matthew, Mark, Luke, }.

21. {House, bungalow, cottage, }.

22. {Mid-off, cover point, square leg, }.

23. {London, Paris, Rome, }.

24. {Alsatian, Pekinese, Dachshund, }.

25. {17, 11, 13, 15, }.

26. {7, 11, 13, 17, }.

27. {4, 9, 16, }.

28. {4, 8, 16, 32, }.

29. $\{\frac{1}{10}, \frac{2}{9}, \frac{3}{8}, \quad \}$.

30. {·14, ·143, ·1429, ·14286, }.

Is it possible or impossible to list *all* the members of the following sets?

31. The names of all the children in your school.

32. The names of all the people who live in England.

33. The names of all the people in the world.

34. All the distances between 1 ft and 2 ft.

35. All the odd numbers between 0 and 1000.

36. All the odd numbers.

37. All the odd numbers between 11 and 13.

38. All the sums of money that can be formed by using some or all of the set of coins {penny, sixpence, shilling, florin, half-crown}.

39. All the sums of money that can be formed by using only pennies and sixpences.

40. All the prime numbers greater than one million.

ELEMENTS OF A SET

So far we have described the objects in the sets we have dealt with as *members*. Mathematicians prefer to speak of them as *elements*.

Examples. (i) Spain is an *element* of the set {Spain, France, Italy}.

We can save time by writing "$\in$" to stand for "is an element of the set", so we could rewrite the previous example as:

$$\text{Spain} \in \{\text{Spain, France, Italy}\}.$$

(ii) To express the fact that something is *not* an element of a certain set we can write $\notin$. Thus

$$\text{Germany} \notin \{\text{Spain, France, Italy}\}.$$

Exercise 1(*b*)

If A is the set $\{a, b, c, d, e\}$, $B = \{p, q, r\}$, $C = \{1, 2, 3\}$, state whether the following statements are *true* or *false*.

1. $a \in A$ **2.** $f \in A$ **3.** $k \notin A$ **4.** $p \in B$

5. $s \in B$ **6.** $a \notin B$ **7.** $0 \in C$ **8.** $1 \in C$

9. $4 \notin C$ **10.** $2 \in$ the set of even numbers.

11. $15 \in$ the set of prime numbers

12. $256 \in$ the set of integral powers of 2.

13. $11 \notin$ the set of natural numbers less than 11.

14. $-7 \in$ the set of real numbers greater than -5.

15. $\{2\} \in C$.

THE EMPTY SET

Sometimes a set contains no elements. We write $\emptyset$ or simply $\{\ \}$ to stand for such a set.

Examples. (i) The set of women in the England cricket XI is $\emptyset$.

(ii) The set of people living in Yorkshire who are 12 ft tall is $\emptyset$.

(iii) The set of natural numbers greater than 5 and less than 6 is $\emptyset$.

Exercise 1(*c*)

In each of the following, list all elements of the set where possible. Otherwise write "impossible to list" or, if there are no elements to list, write $\emptyset$.

1. The proper fractions with denominator 5 between 0 and 1.

2. The proper fractions with numerator 5 between 0 and 1.

3. The prime numbers between 24 and 28.

4. All the shades of colour between yellow and green.

5. All existing animals which are bigger than the elephant.

6. All the numbers between 100 and 150 which are divisible by 7.

7. All the numbers which are divisible by 131.

8. All the numbers which are divisible by 6 and not divisible by 3.

9. All the even square numbers which are not divisible by 4.

10. All the numbers which are divisible by 3 and 7 but not divisible by 21.

EQUAL SETS

Two sets are said to be *equal* if they contain exactly the same elements. The order in which they are listed is unimportant.

Examples. (i) $\{a, b, c\} = \{c, a, b\}$

(ii) $\{5, 10, 17, 26\} = \{n^2+1$ such that n is an integer greater than 1 and less than $6\}$

Exercise 1(*d*)

For each of the following sets, write down the letter of the set in the right-hand column which is equal to it.

1. $\{1, 2, 3, 4\}$	$A = \{1, 2, 3\}$
2. $\{2, 4, 6, 8\}$	$B = \{9\}$
3. $\{p, q, r, s\}$	$C = \{1, 2, 3, 4, 5, 6, 7\}$
4. $\{1, x, x^2\}$	$D = \{5, 6, 7\}$
5. $\{0, 1, 2, 3\}$	$E =$ {the even numbers greater than 1 but less than 10}
6. All the odd numbers greater than 7 but less than 11.	$F = \emptyset$
7. $\{16, 9, 4, 1\}$	$G =$ {the first four natural numbers.}
8. The set of numbers obtained by adding any pair of numbers in the set $\{2, 3, 4\}$.	$H = \{1, 2, 3, 0\}$
	$I = \{x^2, x, 1\}$
	$J = \{q, p, s, r\}$
9. All the factors of 10 which are larger than 10.	$K =$ All the positive integers less than 20 which are perfect squares.
10. All the prime factors of 70.	$L =$ All the digits in the product of 11 and 25.

SUBSETS

Suppose we have two sets, A and $\mathscr{E}$. If every element of the set A is also an element of the set $\mathscr{E}$, we say that A is a subset of $\mathscr{E}$.

Examples. (i) In a school of children ranging in age from 11 yrs to 18 yrs, the set of 13 yrs old children is a subset of the set of all children in the school.

(ii) In a pack of playing cards the set of aces is a subset of all the cards in the pack.

(iii) The set $\{p, q, r\}$ has the following subsets:

$$\{p, q, r\}$$

$$\{p, q\} \quad \{p, r\} \quad \{q, r\} \quad \{p\} \quad \{q\}$$

$$\{r\} \quad \{\ \} \text{ or } \emptyset$$

Notice that the set itself and the empty set may be regarded as subsets. (If we exclude the set itself we have *proper subsets.*)

If A contains the same elements as $\mathscr{E}$ then, of course, we write $A = \mathscr{E}$. If A is a proper subset of $\mathscr{E}$ we express the fact by writing $A \subset \mathscr{E}$. This new symbol stands for the words "is contained in".

Example. The set $\{1, 2, 3\}$ is contained in the set $\{1, 2, 3, 4, 5\}$,

or
$$\{1, 2, 3\} \subset \{1, 2, 3, 4, 5\}$$

Notice that a is an element, whereas $\{a\}$ is a set containing the element a. So while we write

$$a \in \{a, b, c\},$$

we must write
$$\{a\} \subset \{a, b, c\}$$

Exercise 1(*e*)

Supply the connexion between the elements or sets in the left-hand column and the sets in the right-hand column by using whichever of the symbols $\in$, $\notin$, $\subset$, $=$, is appropriate.

1. p	$\{p, q, r\}$
2. $\{p, q, r\}$	$\{q, r, p\}$
3. $\{p\}$	$\{r, p, q\}$
4. s	$\{p, r, q\}$
5. $\{3, 4, 5, 6 \ldots\}$	$\{1, 2, 3, 4, 5, 6 \ldots\}$
6. The positive even integers.	The positive integers.
7. The prime numbers.	The natural numbers.
8. The fourth powers of the natural numbers.	The squares of the natural numbers.
9. The prime numbers.	The numbers of the form $6n \pm 1$ where n is a positive integer.
10. 50	The numbers of the form $n^2 + 1$.

11. Take a *universal set* $\mathscr{E} = \{1, 2, 3, 4, 5, 6, 7, 8, 9\}$. List the subsets whose elements consist of:

(a) All the even numbers in $\mathscr{E}$.
(*b*) All the numbers greater than 5 (i.e. all the numbers > 5).
(c) All the numbers less than 4 (i.e. all the numbers < 4).
(d) All the numbers greater than 9.
(e) All the numbers equal to or greater than 1.

12. Consider a universal set of four children $\mathscr{E}$ = {Andrew, Bill, Cynthia, Diana}. List the subsets which contain:

(a) 3 children
(b) 2 children
(c) 1 boy and 1 girl
(d) Andrew and two others
(e) Diana and one other.

(You may use initial letters A, B, C and D when writing out your answers.)

13. If $\mathscr{E}$ = {Leeds, Blackpool, London, Paris, Scarborough} list all the possible subsets of $\mathscr{E}$ whose elements are:

(a) Capital cities
(b) Seaside towns
(c) Yorkshire towns.

14. If we include the set itself and the empty set, we have seen that a set containing 2 elements has 4 subsets and one containing 3 elements has 8 subsets. How many subsets are there for a set which contains:

(a) 4 elements
(b) 5 elements
(c) n elements?

THE INTERSECTION OF TWO SETS

Suppose we have two sets, A and B. We can form a third set by listing only those elements which are both in A and in B. This third set is called the *intersection* of A and B. We write it as

$$A \cap B$$

Examples. (i) If A = {January, June, July}, B = {June, July, August} then $A \cap B$ = {June, July}.

(ii) If $P = \{a, b, c\}$, $Q = \{c, d, e\}$ then $P \cap Q = \{c\}$.

(iii) If $L = \{l, m, n, p\}$, $M = \{r, s, t\}$ then $L \cap M = \emptyset$.

Exercise 1(f)

1. If $P = \{a, b, c, d\}$, $Q = \{b, c, d, e, f\}$, $R = \{a, c, f\}$ list the following sets:

(a) $P \cap Q$ (b) $Q \cap P$ (c) $P \cap R$
(d) $Q \cap R$ (e) $P \cap (Q \cap R)$ (f) $(P \cap Q) \cap R$

2. Write down the set of months in the year with initial letter J; call this set J. Write down the set of months in the year whose names end with the letter Y; call this set Y. Write down the set of months in the year which have 30 days; call this set S. Call $\mathscr{E}$ the set of all the months in the year. Now list the following sets:

(a) $J \cap Y$ (b) $J \cap S$ (c) $Y \cap S$
(d) $J \cap \mathscr{E}$ (e) $Y \cap \mathscr{E}$ (f) $S \cap \mathscr{E}$
(g) $(J \cap S) \cap Y$ (h) $J \cap (S \cap Y)$

3. X, Y and Z are sets of natural numbers. The elements of X are greater than 1 and less than 6, the elements of Y are greater than 2 and less than 7 and $Z = \{3, 5, 6, 7\}$. (Alternatively we may write: If $x \in X$, $1 < x < 6$; and if $y \in Y$, $2 < y < 7$.)

List the following sets:

(a) $X \cap Y$ (b) $X \cap Z$ (c) $Y \cap Z$
(d) $X \cap X$ (e) $Y \cap Y$ (f) $Z \cap Z$
(g) $X \cap (Y \cap Z)$ (h) $(X \cap Y) \cap Z$

THE UNION OF TWO SETS

Suppose we have two sets, A and B. We can form another set by listing those elements which are either in A or in B or in both A and B. This set is called the *union* of A and B. We write it as $A \cup B$.

Examples. (i) If A = {January, June, July}, B = (June, July, August}, then $A \cup B$ = {January, June, July, August}.

(ii) If $P = \{a, b, c\}$, $Q = \{c, d, e\}$, then $P \cup Q = \{a, b, c, d, e\}$.

Exercise 1(g)

1. Repeat Ex. 1(f), Question 1 but list in this case:

(a) $P \cup Q$ (b) $Q \cup P$ (c) $P \cup R$
(d) $Q \cup R$ (e) $P \cup (Q \cup R)$ (f) $(P \cup Q) \cup R$

2. Repeat Ex. 1(f), Question 2 but list in this case:

(a) $J \cup Y$ (b) $J \cup S$ (c) $Y \cup S$
(d) $J \cup \mathscr{E}$ (e) $Y \cup \mathscr{E}$ (f) $S \cup \mathscr{E}$
(g) $(J \cup S) \cup Y$ (h) $J \cup (S \cup Y)$

3. Repeat Ex. 1(f), Question 3 but list in this case:

(a) $X \cup Y$ (b) $X \cup Z$ (c) $Y \cup Z$
(d) $X \cup X$ (e) $Y \cup Y$ (f) $Z \cup Z$
(g) $X \cup (Y \cup Z)$ (h) $(X \cup Y) \cup Z$

4. If $A = \{1, 2, 3, 4\}$, $B = \{3, 4, 5\}$, $C = \{2, 4, 6\}$ and $\mathscr{E} = \{1, 2, 3, 4, 5, 6\}$, list the following sets:

(a) $A \cup B$ (b) $A \cup C$ (c) $B \cup C$
(d) $A \cup (B \cup C)$ (e) $(A \cup B) \cup C$ (f) $A \cap B$
(g) $A \cap C$ (h) $B \cap C$ (i) $A \cap (B \cap C)$
(j) $(A \cap B) \cap C$ (k) $A \cap (B \cup C)$ (l) $(A \cap B) \cup C$
(m) $A \cup (B \cap C)$ (n) $(A \cup B) \cap C$ (o) $(A \cap B) \cup (A \cap C)$
(p) $(A \cup B) \cap (A \cup C)$ (q) $A \cap \mathscr{E}$ (r) $B \cap \mathscr{E}$
(s) $C \cap \mathscr{E}$ (t) $A \cup \mathscr{E}$ (u) $B \cup \mathscr{E}$
(v) $C \cup \mathscr{E}$ (w) $A \cup A$ (x) $B \cap B$
(y) $A \cap \emptyset$ (z) $A \cup \emptyset$

What do you notice about the answers to (k) and (o); (m) and (p)?

5. Suppose x can only take those values which are elements of the set $\mathscr{E} = \{1, 2, 3, 4, 5, 6, 7, 8, 9\}$. List the set of values of x which satisfy the following:

(a) $x+2 = 4$ (b) $x-1 = 3$ (c) $x > 6$
(d) $x < 5$ (e) $2x = 1$ (f) $x+1 = 12$
(g) $x-1 > 0$ (h) x is prime (i) x is a perfect square number

Write down the intersection of sets (d) and (g). Write down the intersection of sets (a) and (b). Write down the union of sets (a) and (b).

6. Four house captains are given by H or {Alfred, David, Eric Harold}. The tennis six is S or {Brian, Colin, David, Fred, George, Ian}. The rowing eight is R or {Alfred, Brian, Colin, David, Eric, Ian, Harold, John}.

Take the ten boys as a universal set $\mathscr{E}$ and list the following subsets:

(a) $H \cap S$ (b) $S \cap R$ (c) $H \cap R$
(d) $H \cap S \cap R$ (e) $H \cup S \cup R$ (f) $H \cap (S \cup R)$
(g) $(H \cap S) \cup (H \cap R)$ (h) $H \cup (S \cap R)$ (i) $(H \cup S) \cap (H \cup R)$

What do you notice about the answers to (f) and (g); (h) and (i)?

THE COMPLEMENT OF A SET

Suppose we have a class containing boys and girls. Let us call this class the *universal set* $\mathscr{E}$. It may be split (or "partitioned") into two subsets, B the set of boys and G the set of girls.

The *complement* of the set B is that set which consists of all elements in $\mathscr{E}$ which are *not in* B. We denote it by B'. In this case B' is the set of children in the class who are *not* boys, i.e. $B' = G$.

Again, if S is the set of swimmers in the class $\mathscr{E}$, then S' is the set of non-swimmers.

Examples. (i) If $\mathscr{E} = \{a, b, c, d\}$ and $P = \{d\}$ then $P' = \{a, b, c\}$.

(ii) If $\mathscr{E} = \{1, 2, 3, 4, 5, 6\}$ and $P = \{2, 4, 6\}$ then $P' = \{1, 3, 5\}$.

Notice that in each case

$$P \cup P' = \mathscr{E} \text{ and } P \cap P' = \emptyset$$

Exercise 1(*h*)

1. If $\mathscr{E} = \{l, m, n, p, r\}$ list the complements of the following sets:

(a) $\{l, m, n\}$ (b) $\{l\}$ (c) $\{l, m, n, p, r\}$
(d) $\emptyset$ (e) $\{l, m, n\} \cup \{n, p\}$
(f) $\{l, m, n\} \cap \{m, n, p\}$

2. If $\mathscr{E}$ is the set $\{l, m, n, p, r\}$, $P = \{l, m, n, r\}$, $Q = \{l, n, p\}$ list the following sets:

(a) P' (b) Q' (c) $P' \cup Q'$
(d) $P' \cap Q'$ (e) $P \cup Q$ (f) $P \cap Q$
(g) $(P \cup Q)'$ (h) $(P \cap Q)'$ (i) $P \cup P'$
(j) $P \cap P'$ (k) $Q \cup Q'$ (l) $Q \cap Q'$

What do you notice about the answers to (c) and (h); (d) and (g)?

3. Write down the set of months in the year with initial letter J; call this set J. Write down the set of months having 30 days; call this set S. Now list the following sets:

(a) J' (b) S' (c) $(J \cup S)'$
(d) $(J \cap S)'$ (e) $J' \cap S'$ (f) $J' \cup S'$

Which of these sets are equal? Did you expect them to be equal before you worked them out?

4. The days on which John has mathematics form the set M = {Monday, Tuesday, Thursday, Friday} and the days on which he has physics form the set P = {Tuesday, Wednesday, Friday}. $\mathscr{E}$ = {Monday, Tuesday, Wednesday, Thursday, Friday}.

Can you say straight away which of the following sets are equal without actually listing them? If not, list each set and then state which ones are equal.

(a) $M \cup P$ (b) $P \cup M$ (c) $M \cap P$
(d) $P \cap M$ (e) P' (f) M'
(g) $(P \cup M)'$ (h) $P' \cap M'$ (i) $(P \cap M)'$
(j) $P' \cup M'$

Can you express each of the following as a single letter?

(k) $M \cap M$ (l) $M \cup M$ (m) $M \cup (M \cap P)$
(n) $M \cap (M \cup P)$ (o) $M \cap \mathscr{E}$ (p) $P \cup \emptyset$

If further, the days on which John has German form the set G = {Monday, Wednesday, Thursday}, show that:

(q) $M \cap (P \cup G) = (M \cap P) \cup (M \cap G)$
(r) $M \cup (P \cap G) = (M \cup P) \cap (M \cup G)$

VENN DIAGRAMS

The intersections and unions of sets may be represented diagrammatically in the following way. Let A be the set of points on and within the boundary of the circle marked A, and let B be the set of points on and within the boundary of the circle marked B.

In Fig. 1 the sets A and B do *not intersect*. We say that they are *disjoint*. $A \cup B$ is represented by the shaded areas and their boundaries. $A \cap B$, of course, is the empty set.
In Fig. 2 $A \cup B$ is represented by the whole figure and its boundary. $A \cap B$ is represented by the shaded area and its boundary.

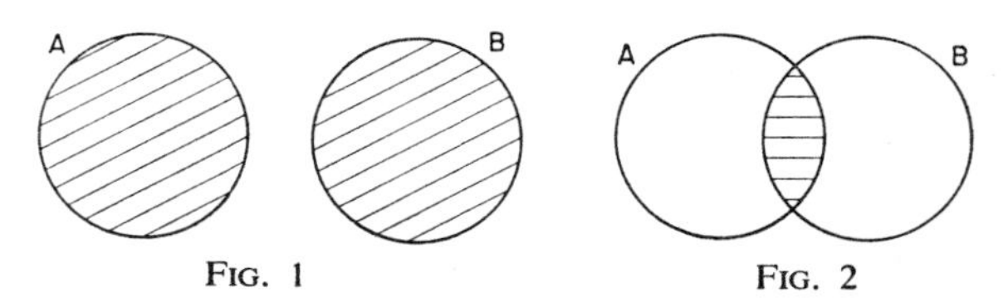

FIG. 1 FIG. 2

Example. (i) Verify by a diagram that

$$A \cap (B \cup C) = (A \cap B) \cup (A \cap C).$$

In Fig. 3 The area shaded horizontally represents $B \cup C$.
The cross-shaded area is clearly $A \cap (B \cup C)$.
In Fig. 4 The area shaded horizontally represents $A \cap B$.
The area shaded vertically represents $A \cap C$.
The total shaded area is therefore $(A \cap B) \cup (A \cap C)$.

Now the cross-shaded area in Fig. 3 and the total shaded area in Fig. 4 are equal.

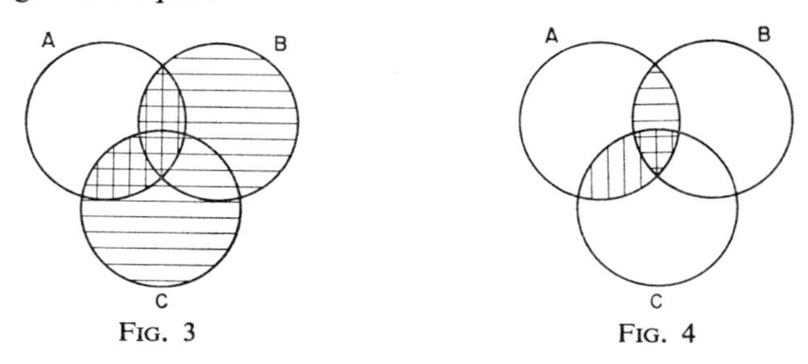

FIG. 3 FIG. 4

Hence $A \cap (B \cup C)$ and $(A \cap B) \cup (A \cap C)$ are represented by the same area. This demonstrates that the two expressions are equal.

Example. (ii) Verify by a diagram that $(A \cap B)' = A' \cup B'$.

In Fig. 5 the shaded area represents $A \cap B$, hence the *unshaded* area represents $(A \cap B)'$.

In Fig. 6 the area shaded vertically represents A' and the area shaded horizontally represents B'.

Hence the total area shaded in any fashion represents $A' \cup B'$. Now this is the same area as the unshaded area in Fig. 5. Hence we conclude that $(A \cap B)' = A' \cup B'$.

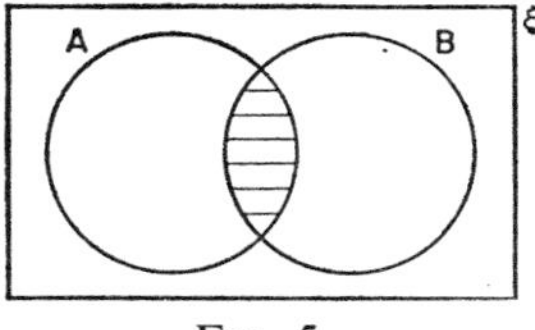

FIG. 5

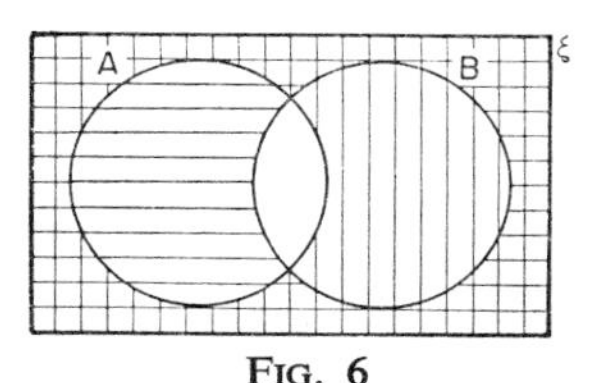

FIG. 6

Exercise 1(*i*)

1. Complete the following in the simplest possible way:

(a) $A \cap \mathscr{E} =$ (b) $A \cup \emptyset =$
(c) $A \cap \emptyset =$ (d) $A \cup \mathscr{E} =$
(e) $A \cap A =$ (f) $A \cup A =$
(g) $A \cap (A \cup B) =$ (h) $A \cup (A \cap B) =$

(Use Venn diagrams where necessary.)

2. Represent each of the following expressions by an appropriately shaded region in a Venn diagram:

(a) $A \cap (B \cup C)$ (b) $(A \cap B) \cup (A \cap C)$
(c) $A \cup (B \cap C)$ (d) $(A \cup B) \cap (A \cup C)$
(e) $(A \cup B)'$ (f) $A' \cap B'$
(g) $(A \cap B)'$ (h) $A' \cup B'$

State which of these expressions are equal.

3. Show by means of diagrams that:

(a) $A \cup (A' \cap B) = A \cup B$ (b) $A \cap (A' \cup B) = A \cap B$

Without further working simplify the expressions:

(c) $B \cup (B' \cap A)$ (d) $B \cap (B' \cup A)$

4. Either by the use of diagrams, or by using any of the results obtained in Questions 1–3, verify the following:

(a) $A' \cup B' \cup C' = (A \cap B \cap C)'$
(b) $A' \cap B' \cap C' = (A \cup B \cup C)'$
(c) $(A \cup B') \cap (A \cup B) = A$
(d) $(A' \cup B') \cap (A' \cup B) = A'$
(e) $(A' \cup B') \cap (A' \cup B) \cap (A \cup B) = A' \cap B$
(f) $(A \cup B' \cup C') \cap [A \cup (B \cap C)] = A$

THE LAWS OF SETS

If you have obtained the correct answers to Questions 1 and 2 in Ex. 1(i) you will have all the rules necessary to manipulate and simplify expressions in the algebra of sets. At a later stage you will find that the algebra of sets may be used to solve logical problems and to design electrical switching circuits. For these purposes our ordinary algebra is no use whatsoever.

In the previous exercise we have obtained the laws of set algebra by using shaded diagrams. This method does not really *prove* that the laws are true. The only satisfactory method, for example, of *proving* that $A \cap (A \cup B) = A$ is to show that *any* element of A is also an element of $A \cap (A \cup B)$ and vice-versa. We can do this as follows:

Let	$x \in A$ (i.e. x is *any* element of the set A)	
then	$x \in A \cup B$.	
But if	$x \in A$ and $x \in A \cup B$	
then	$x \in A \cap (A \cup B)$.	
Hence	$A \subset A \cap (A \cup B)$	(1)

Now if	$x \in A \cap (A \cup B)$	
then clearly	$x \in A$	
i.e.	$A \cap (A \cup B) \subset A$	(2)

From 1 and 2 it follows that $A \cap (A \cup B) = A$.

Exercise 1(*j*)

Prove that the following laws are true in the algebra of sets:

(a) $A \cup (A \cap B) = A$
(b) $A \cup (B \cap C) = (A \cup B) \cap (A \cup C)$
(c) $A \cap (B \cup C) = (A \cap B) \cup (A \cap C)$
(d) $(A \cup B)' = A' \cap B'$
(e) $(A \cap B)' = A' \cup B'$

PROBLEMS

Certain types of problems in arithmetic may be solved quite easily by using Venn diagrams. When we are given information about "overlapping" groups of people or objects we can analyse the information systematically by representing them as intersecting sets and labelling each region with the number of people or objects or "elements" it contains.

Example. (i). In a class of 20 boys, 16 play soccer, 12 play rugby and 2 are not allowed to play games. How many play soccer and rugby as well?

Let S, R represent the sets of boys who play soccer and rugby respectively. Let $n(S)$, $n(R)$ denote the number of boys (or "elements") in the sets S, R respectively, i.e. $n(S) = 16$; $n(R) = 12$.

Now if x boys play both games we may represent the information as shown in Fig. 7.

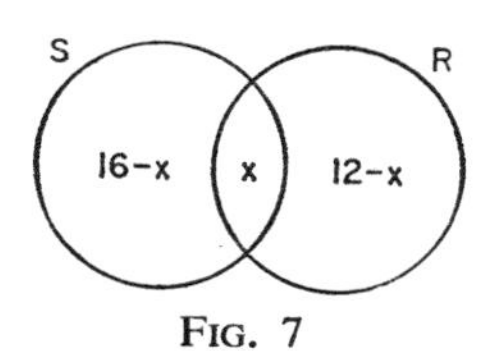

FIG. 7

Then as $\quad n(S \cup R) = 18$
we have $\quad 16-x+x+12-x = 18.$
Therefore $\quad 28-x = 18.$
$\therefore \; x = 10.$

Hence 10 boys play both games.

Example. (ii). 100 vehicles took the Ministry of Transport test and 60 passed. Amongst the remainder, faults in brakes, lights and steering occurred as follows:

brakes only—12
brakes, steering and lights—3
steering and lights only—2.
brakes and steering—5
brakes and lights—8

Equal numbers of cars having only one fault failed because of steering or lighting deficiency. How many cars had faulty lights? How many cars had only one fault?

Represent the sets of vehicles failing on steering, brakes and lights by intersecting circular areas S, B and L respectively.

Let the number of vehicles failing steering or lighting tests only be x in each case. If we assign the numbers given to their respective regions in the Venn diagram this should appear as shown in Fig. 8.

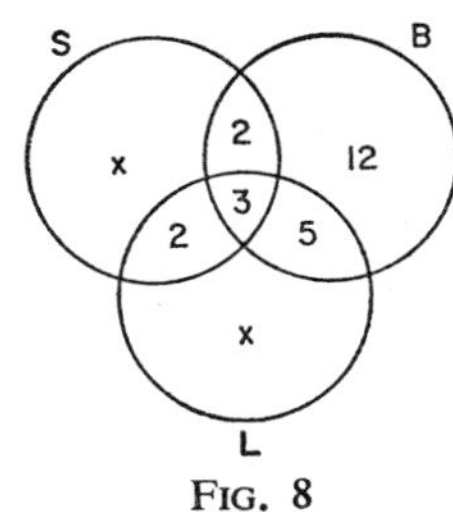

FIG. 8

Now $$n(S \cup B \cup L) = 40$$

i.e. $$2x+24 = 40$$

$$\therefore\ x = 8$$

$$\therefore\ n(L) = 8+2+3+5$$

Hence number of cars with faulty lights = 18.

Further, number of cars having one fault only $= 2x+12$
$= 28$

Exercise 1(k)

1. In a group of 20 children, 8 drink tea but not coffee and 13 like tea. How many drink coffee but not tea?

2. Out of a group of 20 teachers in a school, 10 teach mathematics, 9 teach physics and 7 teach chemistry; 4 teach mathematics and physics but none teaches both mathematics and chemistry. How many teach chemistry and physics? How many teach only physics?

3. The school cricket XI is supposed to wear cricket boots, white flannels and white shirts. During a certain fixture, 7 boys were wearing cricket boots, 9 wore white flannels and 9 wore white shirts. 6 wore white flannels and cricket boots, 6 wore white shirts and cricket boots, and 7 wore white shirts and flannels. How many members of the team were correctly dressed? How many were not wearing cricket boots?

4. Of 50 students, 15 play tennis, 20 play cricket and 20 do athletics. 3 play tennis and cricket, 6 play cricket and do athletics, and 5 play tennis as well as doing athletics. 7 play no games at all. How many play cricket, tennis and do athletics?

5. In the Arts faculty of a certain college an analysis of the courses taken by 70 students in economics, history and geography contained the following figures:

economics only—6
economics but not history—18
economics and geography—36
economics—53
geography—50
history and geography—34

How many took history?
How many took history and geography but not economics?

6. *The Times*, the *Post* and the *Echo* sell in equal numbers to a sample of 497 readers. Analysis of this sample reveals that 26 people read the *Post* and the *Echo*, 50 read *The Times* and the *Echo*, 38 read the *Post* and *The Times* and 11 read all three papers. How many people read only (a) *The Times*, (b) the *Echo*, (c) only one of these papers?

7. At a cultural gathering of 40 people there are 27 men, 20 musicians and 8 singers. 6 of the women are not musicians and 22 of the men are not singers. How many of the women are musicians but not singers?

2
SETS OF POINTS

THE expression $\{x|x > 2\}$ means "the set of values of x for which x is greater than two". If we wish to start listing the elements of this set we must know which values of x we may take. If, for example, x may take values which are positive whole numbers (positive integers) between 0 and 10, i.e. $x \in \mathscr{E}$ where $\mathscr{E} = (1, 2, 3, 4, 5, 6, 7, 8, 9\}$, then clearly $\{x|x > 2\} = \{3, 4, 5, 6, 7, 8, 9\}$. If $x \in \mathscr{E}$, where $\mathscr{E} = \{0, 1, 2\}$, then in this case $\{x|x > 2\} = \emptyset$. If x may take any positive integral value then $\{x|x > 2\} = \{3, 4, 5 \ldots\}$, and we have a set with an infinite number of elements.

Finally, of course, it may be that x can take any real value, and in this case the members of the set crowd so closely together that we cannot even begin to list them in order of size. The first number of the set in this case is of the form 2·0000 . . . 0001, but it is impossible to decide how many noughts we should write. Indeed we could never finish writing them even if we started!

Example. If x may take only values which are positive integers, list the sets (i) $\{x|x > 2\} \cup \{x|x < 3\}$

(ii) $\{x|x > 2\} \cap \{x|x < 5\}$

(iii) $\{x|2x-3 = 0\}$

(i) $\{x|x > 2\} \cup \{x|x < 3\}$

$$= \{3, 4, 5 \ldots\} \cup \{1, 2\}$$
$$= \{1, 2, 3, 4, 5 \ldots\}$$

(ii) $\{x|x > 2\} \cap \{x|x < 5\}$

$$= \{3, 4, 5, 6 \ldots\} \cap \{1, 2, 3, 4\}$$
$$= \{3, 4\}$$

(iii) If $2x-3 = 0$, $x = 3/2$ but since x is restricted to integral values $\{x|2x-3 = 0\} = \emptyset$.

Exercise 2(*a*)

1. If x takes only values which are positive integers, list the following sets:

(a) $\{x|x>4\}$ (b) $\{x|x\geqq 4\}$ (c) $\{x|x<8\}$
(d) $\{x|x>4\} \cap \{x|x<8\}$
(e) $\{x|x>4\} \cup \{x|x<8\}$
(f) $\{x|x<4\} \cap \{x|2<x<5\} \cap \{x|x>3\}$
(g) $\{x|x>4\} \cup \{x|2<x<5\} \cup \{x|x<3\}$
(h) $\{x|x>6\} \cup [\{x|x>5\} \cap \{x|x<7\}]$

2. If $\mathscr{E} = \{x|x$ is a positive integer$\}$, list the subsets of $\mathscr{E}$ which satisfy the following:

(a) $x+4 = 6$ (b) $x+4 = 2$ (c) $2x+8 = 2(x+4)$
(d) $2x+3 = 0$ (e) $x+2 > 3$ (f) $x+6 > 2$
(g) $x+3 < 4$

3. If $x \in \mathscr{E}$ and $\mathscr{E} = \{1, 2, 3, 4, 5, 6, 7, 8, 9\}$, list the subsets of $\mathscr{E}$ which satisfy the following:

(a) x is an even number (b) $5+x > 10$
(c) $5+x < 10$ (d) $3x+1 = 10$ (e) $x+4 = 15$

4. If $x \in \mathscr{E}$ and $\mathscr{E}$ is the set of all the natural numbers, list the following subsets of $\mathscr{E}$:

(a) $\{x|3x-1 = 5\}$ (b) $\{x|x+2 > 5\}$
(c) $\{x|x+6 = 6+x\}$ (d) $\{x|x/2 = 5\}$
(e) $\{x|5x-1 = 2\}$

ORDERED PAIRS

Suppose that x and y are typical elements of two sets X, Y respectively, i.e. $x \in X$ and $y \in Y$. The set (x, y) is then called an *ordered pair*. It is clearly a "pair" of numbers and we call it "ordered" because the order is of great importance. For example the ordered pairs (2, 3) and (3, 2) are not equal. For this reason (x, y) is not really a set and we do not write it in curly set brackets.

To illustrate in diagrammatic form, the elements of X may be marked off on a horizontal axis (the X axis) and those of Y on a vertical line (the Y axis). Thus, if $X = \{1, 2, 3\}$ and $Y = \{1, 2, 3, 4\}$, the set of all possible ordered pairs (x, y) (sometimes called the *product set* $X.Y$ or the *Cartesian product* $X \times Y$) is:

$$\begin{Bmatrix} (1, 4) & (2, 4) & (3, 4) \\ (1, 3) & (2, 3) & (3, 3) \\ (1, 2) & (2, 2) & (3, 2) \\ (1, 1) & (2, 1) & (3, 1) \end{Bmatrix}$$

These are represented by an array of dots (or *lattice*) as shown in Fig. 1.

Example. If $x \in X$, $y \in Y$ where $X = \{1, 2, 3\}$ $Y = \{1, 2, 3, 4\}$, list the following subsets of the product set $X.Y$:

(i) The set of ordered pairs for which $y = x$. This set contains only ordered pairs (x, y) where $y = x$ and is $\{(1, 1)(2, 2)(3, 3)\}$. It may be represented by the set of ringed dots in Fig. 1.

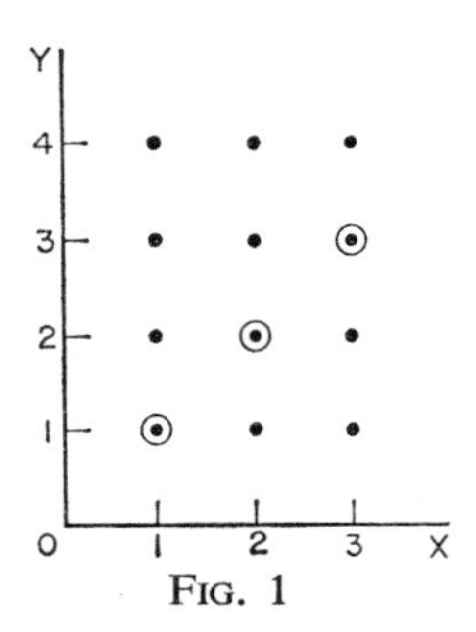

FIG. 1

(ii) $\{(x,y)|x+y = 4\}$ and $\{(x,y)|x+y = 4\} \cap \{(x,y)|y = x\}$. We require the set of ordered pairs (x, y) such that $x+y = 4$, i.e.

$$\{(x, y)|x+y = 4\} = \{(1, 3)(2, 2)(3, 1)\}$$

From 1 $\quad \{(x, y)|y = x\} = \{(1, 1)(2, 2)(3, 3)\}$

$$\therefore \{(x, y)|x+y = 4\} \cap \{(x, y)|y = x\} = \{(2, 2)\}$$

Note that the *intersection* of these two sets is the point (ordered pair) at which the graphs of the lines $x+y = 4$, $y = x$, *intersect*. Alternatively the intersection is the "solution set" of the simultaneous equations $x+y = 4$, $y = x$.

(iii) List the sets $\{(x, y)|y > x\}$; $\{(x, y)|y < x\}$ and $\{(x, y)|y > x\} \cup \{(x, y)|y = x\} \cup \{(x, y)|y < x\}$.

Clearly $\{(x, y)|y > x\} = \{(1, 2)(1, 3)(1, 4)(2, 3)(2, 4)(3, 4)\}$. These ordered pairs may be represented by the dots *above* the line of ringed dots in Fig. 1. Also

$$\{(x, y)|y < x\} = \{(2, 1)(3, 1)(3, 2)\}$$

and these may be represented by all the dots *below* the line of ringed dots in fig. 1. Hence

$$\{(x, y)|y > x\} \cup \{(x, y)|y = x\} \cup \{(x, y)|y < x\}$$

contains all the ordered pairs of the universal product set $X.Y$ and is represented by the whole array of dots. We must surely expect this, since if y may be greater than, equal to, or less than x, there is no restriction at all upon the choice of ordered pairs.

Exercise 2(*b*)

1. x and y take only values which are elements of the set $\{1, 2, 3, 4\}$. List the set of all ordered pairs (x, y). List the subsets which satisfy the following relationships:

(a) $y = x$ (b) $y = x+1$ (c) $y \geqq x$
(d) $y < x$ (e) $x+y = 6$ (f) $x = 2$
(g) $y = 4$ (h) $\{(x, y)|y = x\} \cap \{(x, y)|y = x+1\}$
(i) $\{(x, y)|x = 2\} \cap \{(x, y)|y = 4\}$
(j) $\{(x, y)|y \geqq x\} \cup \{(x, y)|y < x\}$

Represent each of these sets on a lattice diagram.

2. If $x \in X$, $y \in Y$ where $X = \{0,1,2,3,4\}$, $Y = \{0,1,2\ 3,4\}$, list the following subsets of ordered pairs of the product set $X.Y$.

(a) $\{(x, y)|y = 2x\}$ (b) $\{(x, y)|y = x+1\}$
(c) $\{(x, y)|y = x+2\}$ (d) $\{(x, y)|y = -\frac{1}{2}x+4\}$
(e) $\{(x, y)|x+y \leqq 4\}$ (f) $\{(x, y)|x+y \geqq 2\}$

(g) $\{(x, y) | y = 2x\} \cap \{(x, y) | y = x+1\}$
(h) $\{(x, y) | y = 2x\} \cap \{(x, y) | y = x+2\}$
(i) $\{(x, y) | y = x+1\} \cap \{(x, y) | y = -\frac{1}{2}x+4\}$
(j) $\{(x, y) | x+y \geqq 2\} \cap \{(x, y) | x+y \leqq 4\}$

Represent each of these sets on a lattice diagram.

3. If $x \in X$, $y \in Y$ and $X = Y = \{0, 1, 2, 3, 4\}$, list the following subsets of X and of $X.Y$:

(a) $\{x | x-2 = 0\}$
(b) $\{x | x-3 = 0\}$
(c) $\{x | (x-2)(x-3) = 0\}$
(d) $\{x | x+4 = 0\}$
(e) $\{x | x^2-4x = 0\}$
(f) $\{(x, y) | (x-2)(y-3) = 0\}$
(g) $\{(x, y) | x-2 = 0\} \cap \{(x, y) | y-3 = 0\}$
(h) $\{(x, y) | y = \frac{1}{2}x+1\} \cap \{(x, y) | x+y = 4\}$
(i) $\{(x, y) | 2x+y < 4\} \cap \{(x, y) | x+2y < 4\}$
(j) $\{(x, y) | x+y > 6\} \cup \{(x, y) | x+y < 2\}$

Represent the sets (f)–(j) on lattice diagrams.

4. If $x \in X$, $y \in Y$ and $X = Y = \{1, 2, 3, 4\}$, mark in a diagram a lattice of points representing the product set $X.Y$. Using a separate diagram in each case, ring the points which represent the ordered pairs in the following subsets:

(a) $\{(x, y) | y = x\}$
(b) $\{(x, y) | y = x+1\}$
(c) $\{(x, y) | y \geqq x\}$
(d) $\{(x, y) | y < x\}$
(e) $\{(x, y) | x+y = 6\} \cap \{(x, y) | 2x-3y = 2\}$
(f) $\{(x, y) | (x-2)(y-4) = 0\}$
(g) $\{(x, y) | x > 1\} \cap \{(x, y) | x+y \leqq 4\} \cap \{(x, y) | y > 1\}$
(h) $\{(x, y) | x^2+y^2 < 25\}$

LINES AND REGIONS

We now consider what happens to $\{(x, y) | x+y = 4\}$ as we remove our previous restrictions upon the values of x and y.

(i) If $x \in X$, $y \in Y$ and $X = Y =$ the set of integers, positive, negative or zero, then the set $\{(x, y) | x+y = 4\}$ now contains an infinite number of ordered pairs which we may list as $\{\ldots(-3, 7)(-2, 6)(-1, 5)(0, 4)(1, 3)(2, 2)(3, 1)(4, 0)(5, -1)$ $(6, -2)\ldots\}$ and we have an endless straight line of points, some of which we have represented as *unringed* dots in Fig. 2.

(ii) If x and y can now take values which are integers or halves of integers, the set becomes $\{\ldots(-\frac{1}{2}, 4\frac{1}{2})(0, 4)(\frac{1}{2}, 3\frac{1}{2})$ $(1, 3)(1\frac{1}{2}, 2\frac{1}{2})\ldots\}$ and we may now add the ringed dots to our diagram (Fig. 2).

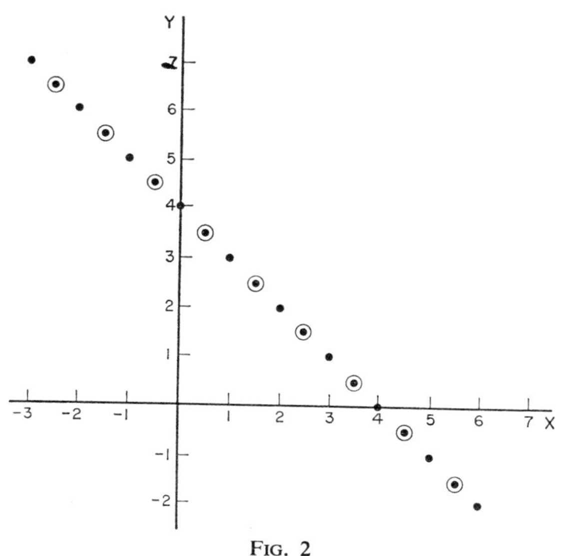

FIG. 2

(iii) If x and y can take *any rational values*, i.e. may be fractions, the line of points becomes very dense indeed. We might almost be tempted to say that the points are now so close that they form a continuous straight line. In fact this is not so.

We have now reached an extremely interesting stage in the study of numbers. If we are given any two unequal rational numbers it is always possible to discover another rational number which lies between them. We construct this third number by adding together the numerators and denominators of the two given numbers. Hence, between $\frac{1}{3}$ and $\frac{1}{2}$ lies the rational number

$\frac{2}{5}$. Again, between $\frac{1}{3}$ and $\frac{2}{5}$ we have $\frac{3}{8}$, and between $\frac{2}{5}$ and $\frac{1}{2}$ we have $\frac{3}{7}$. Repeating this process indefinitely we can fill in the "gap" between $\frac{1}{3}$ and $\frac{1}{2}$ with an infinite number of rational fractions as shown in Fig. 3.

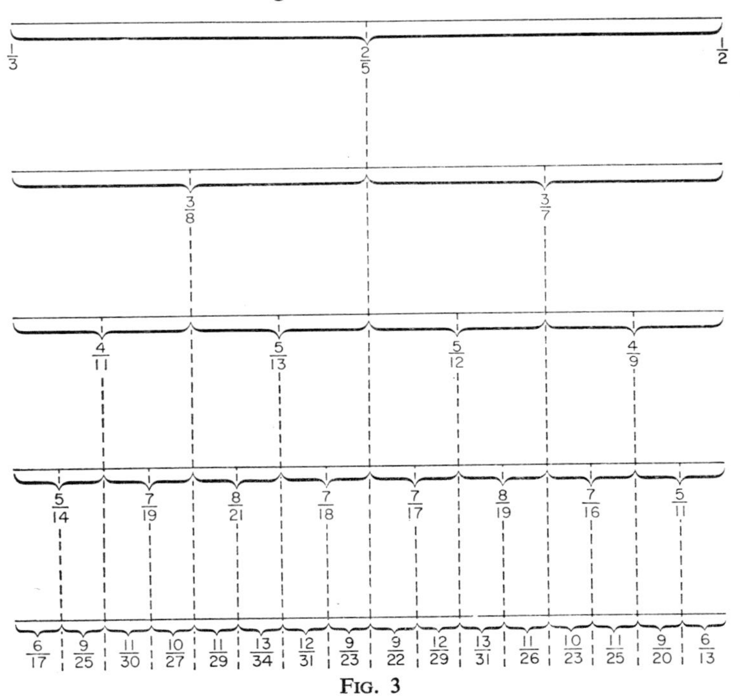

FIG. 3

Curiously enough, however, no matter how far we continue the process, we never succeed in "filling" up the line. An infinite number of "gaps" remain, and the reason for this is that there are values between $\frac{1}{3}$ and $\frac{1}{2}$ which cannot be expressed as rational numbers. These "values", or numbers, are known as *irrational numbers*. An example of such a number is $\sqrt{2}$. An example of a number lying between $\frac{1}{3}$ and $\frac{1}{2}$ is $\frac{1}{4}\sqrt{2}$ or $\cdot 353 \ldots$. This number can only be expressed as a non-terminating, non-recurring decimal fraction. In fact, irrationals occur between *any* pair of unequal rational numbers.

(iv) If x and y may take values which are either integers, rational numbers or irrational numbers, we say that x and y are *real numbers*. From our definition of irrational as "other than rational" real numbers *must* "fill" the line. Thus, $\{(x, y)|x+y = 4\}$ is an infinite set of ordered pairs which form the endless straight line shown in Fig. 4.

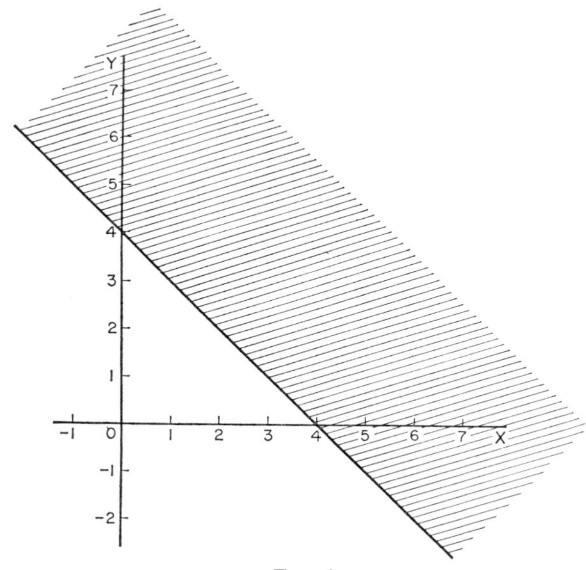

FIG. 4

We call this the *graph* of the equation $x+y = 4$.

In exactly the same way, if x, y are *integers*, then $\{(x, y)|x+y > 4\}$ is represented by a unit lattice of dots above the line $x+y = 4$. But if x, y are real numbers this set of ordered pairs fills the whole plane above $x+y = 4$ and is represented by the shaded area or *region*. Strictly speaking this region does not include its boundary; the boundary itself is only included when we have the set $\{(x, y)|x+y \geqq 4\}$.

Exercise 2(*c*)

Assume throughout this exercise that x and y are both elements of the set of real numbers, i.e. $x \in R$, $y \in R$.

1. Represent graphically by points, lines or shaded regions, the sets of points in Ex. 2(b) Questions 1, 2 and 4.

2. Draw the graphs of the following sets of ordered pairs:

(a) $\{(x, y) | y = x-1\}$ (b) $\{(x, y) | x+y = 4\}$
(c) $\{(x, y) | x = 3\}$ (d) $\{(x, y) | y = 2\}$
(e) $\{(x, y) | xy = 4\}$
(f) $\{(x, y) | y < x^2 \text{ and } -1 < x < 1\}$
(g) $\{(x,y) | x+y < 5\} \cap \{(x,y) | 5y+2x > 10\} \cap \{(x,y) | 2x+y > 6\}$

3. Sketch, on separate diagrams, the graphs of the following equations:

(a) $x+y = 6$ (b) $x+y = 4$
(c) $3x+y = 6$ (d) $x+3y = 6$

4. On the sketch graphs which you have made in Question 3 shade the regions in which the following inequalities are satisfied:

(a) $x+y < 6$ (b) $x+y > 4$ (c) $3x+y > 6$
(d) $x+3y > 6$

5. Draw the graphs (a)–(d) in Question 3 on the same diagram and shade the region in which all the inequalities of Question 4 are satisfied.

6. From the diagram which you have drawn in Question 5 write down the solutions of the following pairs of simultaneous equations:

(a) $x+y = 6$, $3x+y = 6$ (b) $x+y = 6$, $x+3y = 6$ (c) $x+y = 4$, $3x+y = 6$
(d) $x+y = 4$, $x+3y = 6$

7. List the following sets:

(a) $\{(x, y) | x+y = 6\} \cap \{(x, y) | 3x+y = 6\}$

(b) $\{(x, y) | 3x+y = 6\} \cap \{(x, y) | x+3y = 6\}$

(c) $\{(x, y) | x+y = 6\} \cap \{(x, y) | x+3y = 6\}$

(d) $\{(x, y) | x+y = 4\} \cap \{(x, y) | 3x+y = 6\}$

(e) $\{(x, y) | x+y = 4\} \cap \{(x, y) | x+3y = 6\}$

8. Sketch, on the same diagram, the graphs of the following equations:

(a) $y = x+1$ (b) $y = -x+3$ (c) $y = 2x-1$
(d) $y = \frac{1}{2}x+2$

What do you notice about the steepness or *gradient* of these lines and the value of y where they intersect with the Y axis?

9. Use the diagram of Question 8 to solve the following simultaneous equations:

(a) $y = x+1$, $y = 2x-1$ (b) $y = x+1$, $y = \frac{1}{2}x+2$ (c) $y = x+1$, $y = -x+3$

Can you solve these equations (i.e. find the values of x and y which satisfy them) without using a graphical method?

10. In the diagram which you obtained in Question 8, shade the region in which all the following inequalities are satisfied:

$$y > -x+3; \quad y > 2x-1; \quad y < \tfrac{1}{2}x+2$$

Express the set of ordered pairs which may be represented by the points lying in this region as the intersection of three simpler sets.

RELATIONS

The set of all ordered pairs (x, y), where $x \in X$, $y \in Y$, is called the *product set* of X and Y and is denoted $X.Y$. Any proper subset of $X.Y$ is called a *relation* between X and Y. Thus, for example, if $X = \{1, 2\}$, $Y = \{1, 2, 3\}$, then

$$X.Y = \begin{Bmatrix} (1, 3)(2, 3) \\ (1, 2)(2, 2) \\ (1, 1)(2, 1) \end{Bmatrix}$$

Any subsets of this, for example $\{(1,2)(1,3)(2,3)\}$ or $\{(1,1)(2,3)\}$, are *relations*.

The elements need not be *numbers*. If

$$X = \{\text{Alfred's coat, Bill's coat, Claud's coat}\}$$

and $$Y = \{\text{Peg 1, Peg 2, Peg 3}\}$$

or briefly $$X = \{A, B, C\} \qquad Y = \{1, 2, 3\}$$

then
$$X.Y = \begin{Bmatrix} (A, 3)(B, 3)(C, 3) \\ (A, 2)(B, 2)(C, 2) \\ (A, 1)(B, 1)(C, 1) \end{Bmatrix}$$

and any of the following sets are relations:

$$\{(A, 1)(A, 3)(B, 2)(C, 3)\}$$
$$\{(A, 1)(B, 2)(C, 3)\}$$
$$\{(A, 1)(B, 3)(C, 2)\}$$
$$\{(B, 1)(B, 2)\}$$

FUNCTIONS

A relation in which no two different ordered pairs have the same first member is called a function. Thus, in the first example given above, $\{(1, 1)(2, 3)\}$ is a function but $\{(1, 2)(1, 3)(2, 3)\}$ is not, for in the second case two of the ordered pairs, $(1, 2)$ and $(1, 3)$, have the same first member.

Again, in the second case,

$\{(A, 1)(B, 2)(C, 3)\}$ and $\{(A, 1)(B, 3)(C, 2)\}$
are functions, but

$\{(A, 1)(A, 3)(B, 2)(C, 3)\}$ and $\{(B, 1)(B, 2)\}$
are *not* functions.

The second example may help us to see more clearly the difference between a relation and a function. In the function $\{(A, 1)(B, 2)(C, 3)\}$ we are associating, in the ordered pairs, Alfred's coat with peg 1, Bill's coat with peg 2 and Claud's coat with peg 3. This is surely a reasonable thing to do in any school! It is equally reasonable to have $\{(A, 1)(B, 3)(C, 2)\}$, i.e. to have Alfred's coat on peg 1, Bill's coat on peg 3 and Claud's coat on peg 2. In either case, to each boy there corresponds a special peg. We say that between the set of boys' coats and the set of cloakroom pegs there exists a *one: one correspondence.*

Now $\{(A, 1)(B, 1)(C, 2)\}$ is also a function by our definition. Here, perhaps through a shortage of cloakroom space, Alfred and Bill both have to hang their coats on peg 1, while Claud hangs his on peg 2. In this case two elements of one set correspond to one element of the second set and we do not have a one:one correspondence; nevertheless we still have a function. The peg number is still a "function" of the coat which hangs upon it.

Consider, however, the set $\{(A, 1)(A, 3)(B, 2)(C, 3)\}$. Here we have associated Alfred's coat with both peg 1 and peg 3. In practice Alfred will find it difficult to hang his coat on two pegs! This set is a relation but *not* a function.

To take a few examples of the type considered earlier in this chapter we have:

Examples. (i) $\{(x, y) | y = x+4\}$
This is a function.

A listing of its ordered pairs contains no two pairs which have the same first element but different second elements, for to each value of y there is one, and only one, value of x. In fact in this case we also have a one:one correspondence between the set of real number values of x and the set of real y values.

(ii) $\{(x, y) | y > x+1\}$
This is not a function.

A listing of its ordered pairs contains (1, 3) and (1, 4).

(iii) $\{(x, y) | x = 6\}$
This is not a function.

A listing of its ordered pairs contains (6, 1) (6, 2) etc.

(iv) $\{(x, y) | y = 5\}$
This is a function.

No different ordered pairs of the set $\{\ldots (1, 5)(2, 5)(3, 5) \ldots\}$ contain the same first element.

(v) $\{(x, y) | y^2 = x\}$
This is not a function.

A listing of the ordered pairs contains $(1, -1)$ and $(1, 1)$. If, however, we isolate one branch of the graph by writing $\{(x, y) | y^2 = x \text{ and } y \geqq 0\}$ or $\{(x, y) | y = +\sqrt{x}\}$, we do have a function.

Exercise 2(*d*)

1. If $\mathscr{E} = \{1, 2, 3, 4\}$ and $x \in \mathscr{E}$, $y \in \mathscr{E}$, state which of the following relations are functions:

(a) $\{(1, 1)(2, 2)(3, 3)(4, 4)\}$
(b) $\{(1, 2)(2, 2)(3, 2)(4, 2)\}$
(c) $\{(1, 3)(2, 3)(1, 4)(2, 4)\}$
(d) $\{(3, 1)(3, 2)(3, 3)(3, 4)\}$
(e) $\{(4, 1)(3, 2)(2, 3)\}$

2. Which of the following relations are functions?

(a) {(Albert, London)(Ben, Manchester)(Cyril, Manchester)}
(b) {(Albert, London)(Ben, Manchester)(Cyril, Newcastle)}
(c) {(Albert, London)(Ben, Newcastle)(Ben, Manchester)}

Unless otherwise stated, assume in the following questions that $x, y \in R$. State whether the following sets of ordered pairs are relations or functions.

3. $\{(x, y) | x+y = 4\}$
4. $\{(x, y) | x+y > 4\}$
5. $\{(x, y) | y-4 = 0\}$
6. $\{(x, y) | x+2 = 0\}$
7. $\{(x, y) | y = x^2\}$
8. $\{(x, y) | xy = 4\}$
9. $\{(x, y) | x^2+y^2 = 25\}$
10. $\{(x, y) | x^2+y^2 = 25 \text{ and } x > 0\}$
11. $\{(x, y) | x^2+y^2 = 25 \text{ and } y < 0\}$

12. $\{(x, y) | x^2 + y^2 < 25\}$

13. $\{(x, y) | x + y > 4\} \cap \{(x, y) | x + y < 6\}$

14. $\{(x, y) | y - 1 = 0\} \cup \{(x, y) | y - 2 = 0\}$

15. $\{(x, y) | y^2 = x^2\}$

16. $\{(x, y) | y^3 = x\}$

17. $\{(x, y) | y^4 = x\}$

18. $\{(x, y) | x$ is a prime factor of $y\}$

19. $\{(x, y) | y = 1$ if x is odd; $y = 0$ if x is even$\}$ where x, y are elements of the set of integers.

20. $\{(x, y) | x = |y|\}$ (Note: $|y|$ is the numerical value of y irrespective of sign, e.g. $|-2| = 2$.)

21. $\{(x, y) | y = |x|\}$

22. $\{(x, y) | y$ is the father of $x\}$

23. $\{(x, y) | y$ is the son of $x\}$

24. $\{(x, y) |$ the difference between y and x is constant$\}$

25. $\{(x, y) |$ the sum of x and y is an integral multiple of $x\}$

MAPPINGS

A mapping is a function. No essentially new idea is involved However, as the term implies, it is a more graphic way of looking at the idea of a function.

To take a very obvious illustration of the term, consider the set of houses on a building site. We can make a site plan, or *map*, by drawing small squares on a sheet of drawing paper in such a way that these correspond in relative position to the houses on the site. When the plan is complete we have mapped every element of the set of houses H on to the set of squares S. To each house there corresponds one, and only one, square and vice versa. The subset of $H.S$ which constitutes this mapping of H onto S will contain no two different ordered pairs with the same first element, i.e. no house will be represented by two different squares. Thus we have a *function*. The "function" in this case is simply the act of representing a house by a small square on paper.

To indicate that elements of H are mapped on to elements of S by means of a function we write

$$H \xrightarrow{f} S$$

or

$$f: H \to S.$$

If $h \in H$ and $s \in S$ and the house h maps onto the square s we also write $f(h) = s$. Also we say that s is the *image* of h under the mapping f.

In this particular case the mapping is a very special one. As every element of H is mapped on to a unique element of S and the sets contain equal numbers of elements, we call this a *one: one mapping of H on to S.* Further, we sometimes describe s as *the image of h in the set S.*

If we had a map of all the houses in the town, including those on the building site, we should still have a one:one mapping, but this time it would be a mapping of one set *into* a larger one.

In order to have a function it is not necessary for the mapping to be one:one, or indeed that it should be *on to*.

Examples. (i) Let us recall the example of a function $\{(A, 1)(B, 1)(C, 2)\}$ discussed earlier in the chapter.

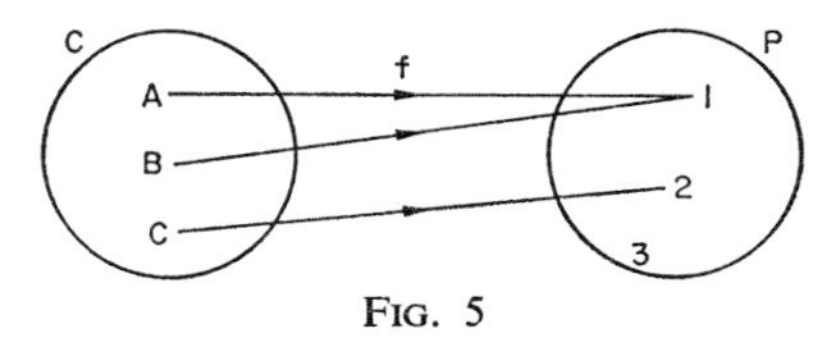

FIG. 5

Here we have a mapping of the elements of a set of coats C *into* the elements of a set of pegs P. The mapping is not one:one.

(ii) $\{(x, y) \mid y = 1\}$ is a function. A listing of the ordered pairs contains $\{\ldots(-2, 1)(-1, 1)(0, 1)(1, 1)(2, 1)\ldots\}$, and no two different ordered pairs have the same first member. In this case we have a mapping of all the real numbers on to the set containing the single element 1.

(iii) $\{(x, y) | y = x^2\}$. This is a function and we may write

$$X \xrightarrow{f} Y,$$

or

$$f: X \to Y,$$

or

$$f(x) = y,$$

where f is the operation of *squaring*. Or, to avoid saying what f does, we may simply write $f(x) = x^2$. In this case the set of positive and negative real numbers is mapped into the set of

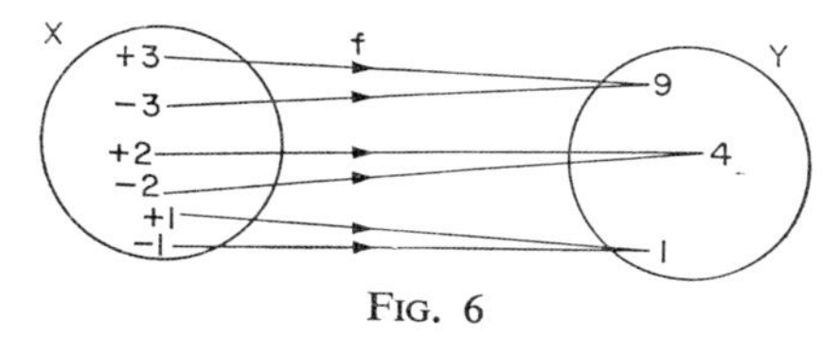

FIG. 6

positive real numbers. It is a two:one mapping. We may also write

$$f(3) = 9, \quad f(2) = 4, \quad f(1) = 1, \quad f(\tfrac{1}{2}) = \tfrac{1}{4}, \quad \text{etc.}$$

SUCCESSIVE MAPPINGS

We sometimes map a set X into or on to a set Y, and then map the set Y into another set Z.

Suppose that $f(x) = x^2$ and $g(y) = y+1$. The mapping of some of the elements may be shown as follows:

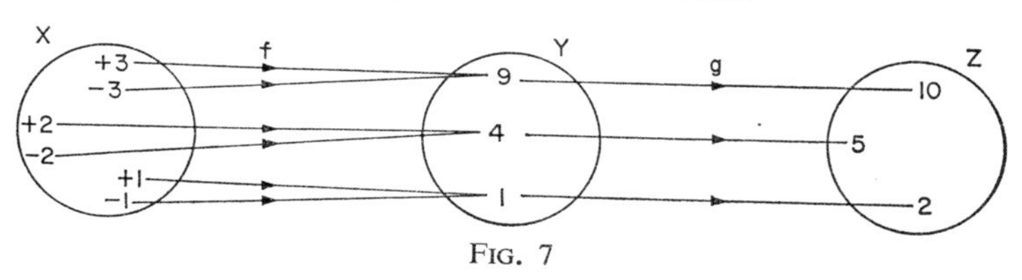

FIG. 7

Thus $\qquad f(3) = 9 \quad \text{and} \quad g(9) = 10$

i.e. $\qquad g[f(3)] = 10 \quad \text{or} \quad g \circ f(3) = 10.$

(or "g of f three equals ten")

Note, however, that

$$g(3) = 4 \quad \text{and} \quad f(4) = 16.$$
$$\therefore\ f[g(3)] = 16 \quad \text{or} \quad f\circ g(3) = 16$$

i.e. $g[f(x)]$ is not the same as $f[g(x)]$. Generally, $g[f(x)] = x^2+1$, but $f[g(x)] = (x+1)^2$.

THE DOMAIN AND RANGE OF A FUNCTION

The *domain* of a function (or a relation) is the set of all the first members of the ordered pairs in the function. The *range* is the set of all the second members of the ordered pairs in the function.

Referring to the examples in the previous section we have:

Example (i). $\{(A, 1)(B, 1)(C, 2)\}$; the domain $= \{A, B, C\}$; the range $= \{1, 2\}$.

Example (ii). $\{(x, y)|y = 1\}$; the domain is the set of real numbers; the range is the element 1.

Example (iii). $\{(x, y)|y = x^2\}$; the domain is the set of real numbers; the range is the set of positive real numbers (including 0).

Exercise 2(*e*)

Assume, unless otherwise stated, that $x \in X$, $y \in Y$, $z \in Z$, and that x, y, z take real values.

In each of the following functions (Questions 1–10) state whether or not the mapping is one:one. Give the domain and range in each case.

1. $\{(x, y)|x+y = 4\}$

2. $\{(x, y)|y-4 = 0\}$

3. $\{(x, y)|y^3 = x\}$

4. $\{(x, y)|y = x^4\}$

5. $\{(x, y)|xy = 4\}$ (Exclude the cases where either $x = 0$ or $y = 0$.)

6. $\{(x, y) | y = 1 \text{ if } x \text{ is odd, } y = 0 \text{ if } x \text{ is even}\}$ where x, y are elements of the set of integers.

7. $\{(x, y) | y = |x|\}$

8. $\{(x, y) | y = x^2+1\}$

9. $\{(x, y) | y = +\sqrt{4-x^2}\}$

10. $\left\{(x, y) | y = \dfrac{x^2}{x^2+1}\right\}$

Two mappings, f and g, are defined as shown in Fig. 8. In questions 11–15:

11. Are the mappings f and g one:one?

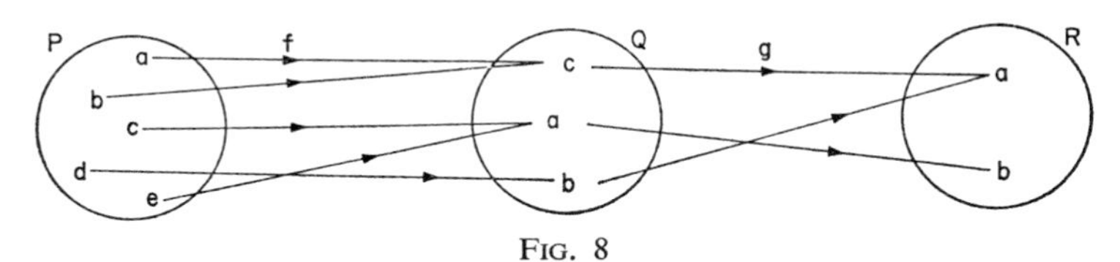

FIG. 8

Write down the values of the following:

12. (a) $f(a)$ (b) $f(b)$ (c) $f(d)$
(d) $f(e)$

13. (a) $g(c)$ (b) $g(a)$ (c) $g(b)$

14. (a) $g \circ f(a)$ (b) $g \circ f(b)$ (c) $g \circ f(c)$
(d) $g \circ f(d)$ (e) $g \circ f(e)$

15. (a) $f \circ g(a)$ (b) $f \circ g(b)$ (c) $f \circ g(c)$

Two mappings, f and g, are defined as shown in Fig. 9. In questions 16–20:

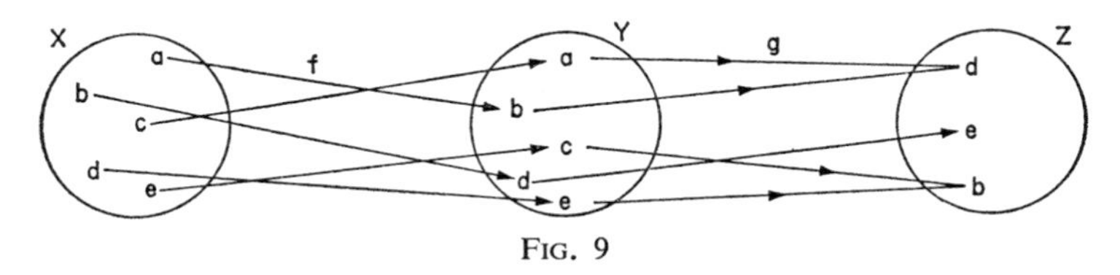

FIG. 9

16. Are either of the mappings one:one?

Write down the values of the following:

17. (a) $f(a)$ (b) $f(b)$ (c) $f(c)$
(d) $f(d)$ (e) $f(e)$

18. (a) $g(a)$ (b) $g(b)$ (c) $g(c)$
(d) $g(d)$ (e) $g(e)$

19. (a) $f \circ g(a)$ (b) $f \circ g(b)$ (c) $f \circ g(c)$
(d) $f \circ g(d)$ (e) $f \circ g(e)$

20. (a) $g \circ f(a)$ (b) $g \circ f(b)$ (c) $g \circ f(c)$
(d) $g \circ f(d)$ (e) $g \circ f(e)$

21. Let the domain of x be the set $\{0, 1\}$. Which of the following functions on this particular domain are equal?

(a) $f(x) = x^2;\ \ g(x) = x^3$
(b) $f(x) = x+1;\ \ g(x) = 3x^2-2x+1$
(c) $f(x) = 2;\ \ g(x) = x^2-x+2$
(*d*) $f(x) = 3x;\ \ g(x) = 2x+1$

22. Find the function which maps any set on to itself.

23. If $f(x) = x+1$ and $g(x) = \dfrac{1}{x+1}$ and the domain of x is $X = \{1, 2, 3, 4\}$, state the ranges of $f(x)$ and $g(x)$.

24. Suppose that x can take any real value except $x = 1$, and that $f(x) = \dfrac{1}{1-x};\ \ g(x) = \dfrac{x-1}{x}$. Evaluate:

(a) $f(-3);\ \ f(-2);\ \ f(2)$.
(b) $g(\tfrac{1}{4});\ \ g(\tfrac{1}{3});\ \ g(-1)$.

Show that for all values of x; $f \circ g(x) = x$.
Is it true to say that $g \circ f(x) = x$?
[If $f \circ g(x) = x$, f is called the *inverse* of the function g. Whatever effect is created by the operation of a function is cancelled by the subsequent operation of its *inverse*.]

25. If $f(x) = x^2$ and $g(x) = \tan x$, write the functions $f \circ g(x)$ and $g \circ f(x)$.

26. If $f_1(x) = x$, $f_2(x) = -x$, $f_3(x) = \frac{1}{x}$, $f_4(x) = \frac{-1}{x}$, what are:

(a) $f_2 \circ f_3(x)$ (b) $f_2 \circ f_4(x)$ (c) $f_3 \circ f_3(x)$
(d) $f_1 \circ f_4(x)$ (e) $f_4 \circ f_4(x)$ (f) $f_2 \circ f_3 \circ f_4(x)$

27. If $f_1(x) = x$, $f_2(x) = \frac{1}{1-x}$, $f_3(x) = \frac{x-1}{x}$, $f_4(x) = \frac{1}{x}$, $f_5(x) = 1-x$, $f_6(x) = \frac{x}{x-1}$, what are:

(a) $f_1 \circ f_2(x)$ (b) $f_2 \circ f_1(x)$ (c) $f_2 \circ f_3(x)$
(d) $f_3 \circ f_2(x)$ (e) $f_3 \circ f_4(x)$ (f) $f_4 \circ f_3(x)$
(g) $f_5 \circ f_5(x)$ (h) $f_4 \circ f_4(x)$

3

LINEAR PROGRAMMING

IN THE last chapter we saw that the set of ordered pairs of real numbers which satisfy an inequality such as $x+y < 4$ may be represented by a shaded region of the x, y plane. In some of the exercises we shaded the region in which several such inequalities were satisfied. This particular type of exercise is of very great practical importance in solving a certain class of problems which arise frequently in industry today.

In the highly competitive markets of the modern world, industrial and commercial enterprises are obliged to ensure that within their organizations factors such as productivity and profit are as large as possible, while wastage, overheads and transport costs are kept as low as possible. The individual conditions which have to be obeyed are often simple and *linear* in form, but there are usually a great many of them. In real life these problems are so complicated that the calculations have to be programmed for a computer. In this chapter we shall look at some problems of this type which have been simplified sufficiently for us to solve them by graphical methods.

Example (i). A bicycle manufacturer makes two models, a sports cycle and a racing machine. In order to make a sports model, 6 man-hours are needed, while a racing model requires 10 man-hours. (By 6 man-hours we mean either 1 man working for 6 hours, or 2 men working for 3 hours, or 3 men working for 2 hours, or 6 men working for 1 hour, etc.) The manufacturer can employ no more than 15 men and these men work 8 hours per day for 5 days each week. The cost of materials amounts to £5 per cycle and the manufacturer's total weekly quota of such materials may not exceed £400. The firm has a contract to supply at least 30 sports models and 20 racers per week. How many cycles of each type should be made in order to obtain the maximum possible profit if (a) the profit on each sports cycle is £1 and on each racing model it is £3, (b) the profit on each sports model is £3 10*s*. 0*d*. and on each racing model it is £4 10*s*. 0*d*.?

Suppose that x sports models and y racing models are made each week, then the total man-hours required per week is $6x+10y$. Now the total number of man-hours available cannot exceed $15\times5\times8$ or 600 hours, i.e.

$$3x+5y \leqq 300 \tag{1}$$

The total weekly cost of materials is £$(5x+5y)$. Since this may not exceed £400 we have:

$$x+y \leqq 80 \tag{2}$$

But at least 30 sports models and 20 racers must be made.

$$\therefore\ x \geqq 30 \tag{3}$$

and

$$y \geqq 20 \tag{4}$$

We now draw, on the same diagram, the graphs of the equations $3x+5y = 300$, $x+y = 80$, $x = 30$ and $y = 20$. The shaded area in Fig. 1 now represents the region in which the inequalities 1, 2, 3 and 4 are all satisfied.

(a) If £P is the total weekly profit, then in this case

$$x+3y = P \tag{5}$$

Now equation (5) may be written as:

$$y = -\frac{1}{3}x + \frac{P}{3}$$

The graphs of this equation for various values of P are a family of parallel straight lines of gradient $-\frac{1}{3}$. Now the solution must lie inside or on the boundary of the shaded area in order to satisfy the requirements 1–4. If, however, P is to be a *maximum*, we require that member of the family $y = -\frac{1}{3}x+\frac{P}{3}$ which makes the *greatest* intercept on the y axis and yet still passes through at least one point of the shaded area. The line required is shown by a dotted line passing through A in Fig. 1. Hence, the required solution is given by $x = 30$, $y = 42$. That is to say the manufacturer must make 30 sports models and 42 racing machines each week. In this case his total weekly profit will be £$(30+42\times 3)$, i.e. £156.

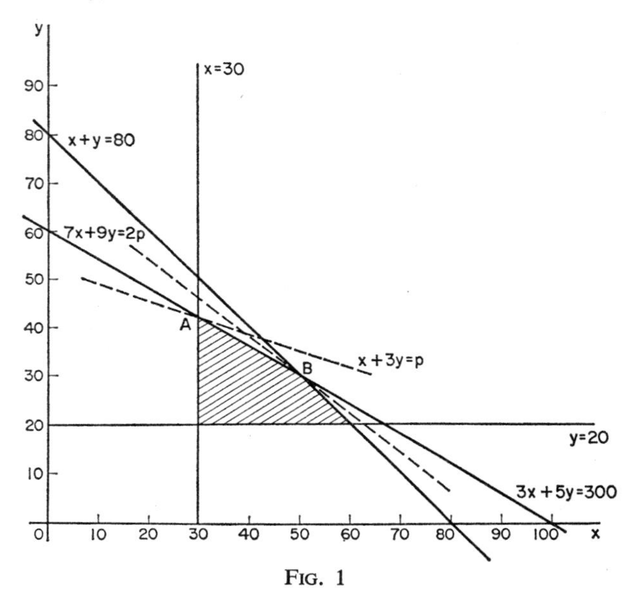

FIG. 1

(b) If £P is the weekly profit, then in this case

$$3\tfrac{1}{2}x+4\tfrac{1}{2}y = P \tag{6}$$

Equation (6) may be written as:

$$y = -\frac{7}{9}x+\frac{2P}{9}$$

The member of this family of straight lines of gradient $-\dfrac{7}{9}$ which passes through at least one point of the shaded area and at the same time makes the greatest possible intercept on the Y axis (i.e. makes $\dfrac{2P}{9}$ a maximum) is shown by the other dotted line passing through B in Fig.1. In this case the required solution is given by $x = 50$, $y = 30$. That is to say, the manufacturer should make 50 sports models and 30 racing machines for a gross weekly profit of £$(3\frac{1}{2} \times 50 + 4\frac{1}{2} \times 30)$ or £310.

You should check that no other ordered pair of values in the shaded area gives so great a profit as this.

Example (ii). A coal merchant has two depots, D_1 and D_2, which contain current stocks 30 tons and 15 tons of coal respectively. Three industrial customers, C_1, C_2 and C_3, place orders for 20 tons, 15 tons and 10 tons of coal respectively. The distances between the various depots and customers are given by the following table:

	C_1	C_2	C_3
D_1	7	4	2
D_2	3	2	2

(distances in miles.)

Assuming that the cost of transporting the coal is a fixed amount per ton per mile, decide from which depots the orders should be distributed in order to keep these costs to a minimum.

Let us suppose that we send x tons of coal from D_1 to C_1 and y tons from D_1 to C_2; then the other deliveries will be as shown below:

ORDERS

		20 *tons* C_1	15 *tons* C_2	10 *tons* C_3
Stocks	30 tons D_1	x	y	$30 - (x + y)$
	15 tons D_2	$20 - x$	$15 - y$	$(x + y) - 20$

Now it is impossible to deliver a negative amount of coal, therefore each of these quantities must be positive, i.e.

$$x+y \leqq 30 \qquad (1)$$
$$x+y \geqq 20 \qquad (2)$$
$$y \leqq 15 \qquad (3)$$
$$x \leqq 20 \qquad (4)$$
$$y \geqq 0 \qquad (5)$$
$$x \geqq 0 \qquad (6)$$

If the total transport cost is T units of money, then we have:

$$T = 7x+4y+2[30-(x+y)]+3(20-x)+2(15-y)+2[(x+y)-20]$$

$$\therefore\ T = 4x+2y+110$$

or

$$y = -2x+\frac{T-110}{2} \qquad (7)$$

[T is, of course, the number of ton-miles, and we obtain the expression by multiplying each delivery by the distance it travels and adding together the results.]

As in the previous example we shade the area in which the inequalities 1–6 are satisfied. In this problem, however, we wish to *minimize* the cost of transport. We require, therefore, that

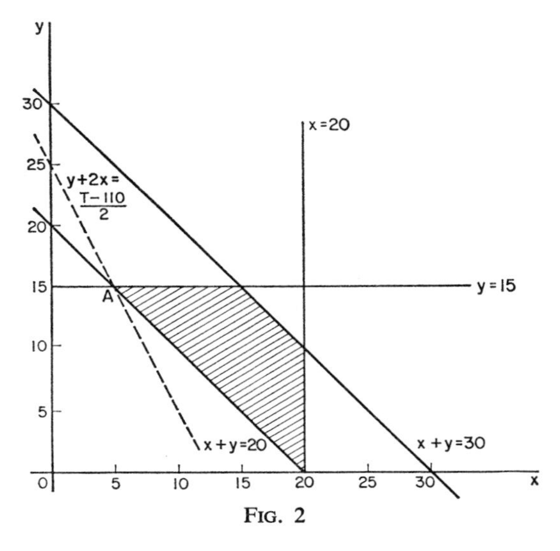

FIG. 2

member of the family of straight lines given by equation (7) which has the smallest possible intercept on the y axis and still passes through at least one point of the shaded region in which the inequalities 1–6 are satisfied. The dotted line shown in Fig. 2 is the one required.

From this we see that the values of x and y for which T has a minimum value are given by the point A where $x = 5$

and $y = 15$. Hence the minimum-cost delivery schedule is as follows:

	C_1	C_2	C_3
D_1	5	15	10
D_2	15	0	0

and the actual minimum cost is $(4 \times 5 + 2 \times 15 + 110)$, or 160 units of money.

Alternatively, the intercept on the y axis is:

$$\frac{T-110}{2} \text{ or } 25$$

i.e.

$$T = 160 \text{ as before.}$$

Exercise 3(*a*)

1. A rectangular board has to be constructed so that its perimeter is less than 20 ft. and greater than 12 ft. The ratio of the adjacent sides must be greater than 1:1 and less than 2:1. What integral dimensions satisfy these requirements?

2. A farmer wishes to buy a number of cows and sheep. Cows cost £18 each and sheep cost £12 each. The farmer has accommodation for not more than 20 animals and cannot afford to pay more than £288. If his reasonable expectation of profit is £11 per cow and £9 per sheep, how many of each should he purchase in order to make his total profit as large as possible?

3. A cement manufacturer has two depots, D_1 and D_2, which contain current stocks 30 tons and 15 tons of cement respectively. Three customers, C_1, C_2 and C_3, place orders for 20 tons, 15 tons and 10 tons of cement respectively. The distances between the various depots and customers are given by the following table:

	C_1	C_2	C_3
D_1	3	4	1
D_2	2	1	5

Assuming that the cost of transporting the cement is a fixed amount per ton per mile, decide from which depots the orders should be distributed in order to make these costs a minimum.

4. A firm which supplies raw material in bulk has two depots, D_1 and D_2, in a certain area. D_1 and D_2 currently stock 140 and 40 tons of material respectively. Two customers, C_1 and C_2, place orders for 100 and 50 tons respectively. The cost of transport is a fixed amount per ton per mile. C_1 is 60 miles from D_1 and 30 miles from D_2; C_2 is 80 miles from D_1 and 20 miles from D_2. How should the deliveries be made in order to minimize transport costs?

(*Hint.* Let C_1, C_2 receive x, y tons respectively from D_1.)

5. 10 gm of alloy A contain 2 gm of copper, 1 gm of zinc and 1 gm of lead; 10 gm of alloy B contain 1 gm of copper, 1 gm of zinc and 3 gm of lead. It is required to produce a mixture of these alloys which contains at least 10 gm of copper, 8 gm of zinc and 12 gm of lead. Alloy B costs $1\frac{1}{2}$ times as much per kilogram as alloy A. Find the amounts of alloy A and alloy B which must be mixed in order to satisfy these conditions in the cheapest possible way.

(*Hint.* Consider a mixture of x, y dekagrams of A and B respectively.)

6. A small manufacturer employs 5 skilled men and 10 semi-skilled men and makes an article in two qualities, a de luxe model and an ordinary model. The making of a de luxe model requires 2 hrs work by a skilled man and 2 hrs work by a semi-skilled man; the ordinary model requires 1 hrs work by a skilled man and 3 hrs work by a semi-skilled man. By union rules no man may work more than 8 hrs per day. The manufacturer's clear profit on the de luxe model is 10/- and on the ordinary model 8/-. How many of each type should he make in order to maximize his total daily profit?

(*Hint.* Assume that he makes x ordinary and y de luxe models each day.)

7. A master printer employs journeymen and apprentices and his facilities are such that he cannot employ more than 9 people altogether. His orders oblige him to maintain an output of at least 30 units of printing work per day. On the average a journeyman does 5 units of printing work and an apprentice 3 units of printing work daily. The Apprentices Act demands that the printer should employ not more than 5 men to 1 apprentice. The Journeymen's union, however, forbids him to employ less than 2 men to each apprentice. How many journeymen and apprentices should he employ? Of the possible results, which is the wisest choice if he has to pay journeymen £2 per day and apprentices £1 per day? If he charges £1 per printing unit to his customers and can sell all his output in excess of 30 units, in which case is his profit a maximum?

8. A dictator seizes power in a small state and proceeds to plan the economy and labour forces. He discovers that there are two motor corporations. Each factory owned by corporation *A* makes each week 30 vans, 10 saloon cars and 10 lorries, while each factory in corporation *B* makes 10 vans, 10 saloon cars and 40 lorries weekly. He finds that the average combined home and overseas market for these vehicles is at least 100 vans, 60 saloon cars and 120 lorries weekly, but that previously these demands have been greatly exceeded in some respects and not met in others. In his reorganization, how many factories in each corporation should continue to operate?

If the labour force in each factory of *A* is half that in each factory of *B*, and if profits are directly proportional to labour force, which answer is best

(a) for economizing on labour?
(b) for maximizing profits?

9. A dietician wishes to mix together two kinds of food so that the vitamin content of the mixture is at least:

Vit. A—9 units; Vit. B—7 units;
Vit. C—10 units; Vit. D—12 units.

The vitamin content of each pound of food is shown below:

	Vit. A	*Vit. B*	*Vit. C*	*Vit. D*
Food 1	2	1	1	1
Food 2	1	1	2	3

Food 1 costs 5/- per lb.
Food 2 costs 7/- per lb.

Find the minimum cost of such a mixture.

10. A mine manager has contracts to supply weekly 1000 tons of grade 1 coal, 700 tons grade 2, 2000 tons grade 3, 4500 tons grade 4. Seams A and seam B are being worked at a cost of £4,000, £10,000 respectively per shift, and the yield in tons per shift from each seam is given by the table below:

	Grade 1	*Grade 2*	*Grade 3*	*Grade 4*
Seam A	200	100	200	400
Seam B	100	100	500	1,500

How many shifts per week should each seam be worked in order to fulfil the contracts most economically?

11. A distributor of certain manufactured articles (units) supplies orders from two depots D_1 and D_2. At a certain time D_1 and D_2 have in stock 80 and 20 of these articles respectively. Two customers, C_1 and C_2, place orders for 50 and 30 units respectively. The transport cost of any article is directly proportional to the distance it is conveyed, and the distance in miles between the depots and the customers is given by the table below:

	C_1	C_2
D_1	40	30
D_2	10	20

From which depots should the orders be despatched in order to minimize the transport costs?

4

SETS, LOGIC AND SWITCHING CIRCUITS

PREMISES, CONCLUSIONS AND VENN DIAGRAMS

THE study of sets has some surprising applications to logical arguments and electrical circuits. We will start by considering how the laws of sets and the use of Venn diagrams can help us to analyse certain types of argument. Consider the argument:

All squares are rectangles. (1)
All rectangles are parallelograms. (2)
Therefore all squares are parallelograms. (3)

The statements (1) and (2) are called *premises*, or *hypotheses*, and (3) is called the *conclusion*. In this case the conclusion quite clearly follows from the premises and we say that the argument is *valid*. Quite briefly we may say that the whole statement comprising (1), (2) and (3) has the value T (true). Although diagrammatic assistance is unnecessary in this case, we may, nevertheless, represent the structure of this argument by the type of Venn diagram shown in Fig. 1, where S is the set of all squares, R is the set of all rectangles and P is the set of all parallelograms.

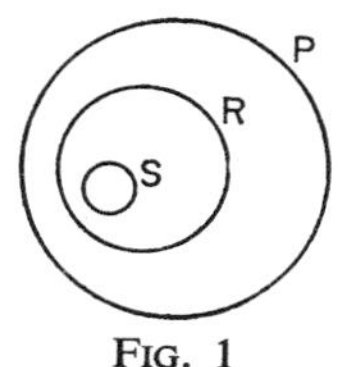

FIG. 1

S is a special subset of R and R is a special subset of P, i.e. $S \subset R$ and $R \subset P$. It follows that $S \subset P$, which is a "shorthand" way of writing the conclusion 3.

In mathematics we may have an argument which is valid but in which the conclusion is untrue. For example:

London is in Ohio.
Ohio is in America.
Therefore London is in America.

This is a perfectly valid argument but the conclusion is untrue because the first premise is untrue.

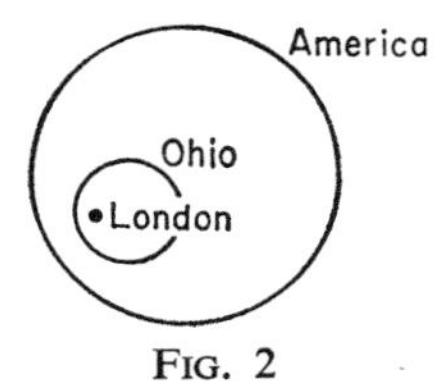

FIG. 2

We sometimes use this type of argument in school mathematics. For example, to prove that the tangent to a circle is perpendicular to the radius drawn to its point of contact, we may commence the proof by assuming that the tangent is not perpendicular to the radius. Using a perfectly correct argument we then come to the conclusion that the tangent must intersect the circle in two points. Since this conclusion is obviously untrue by the definition of a tangent, it follows that our original premise was equally untrue. In other words the tangent is perpendicular to the radius drawn to the point of contact.

On the other hand we sometimes encounter arguments in which a true conclusion is reached as a result of an invalid argument. Consider the statements:

Some problems are mathematical.
Some problems are difficult.
Therefore some mathematical problems are difficult.

Surely no one would object to any of these statements. The argument itself, however, is quite *invalid.*

If M is the set of mathematical problems, D is the set of difficult problems and P is the set of all problems, we see in Fig. 3 that the

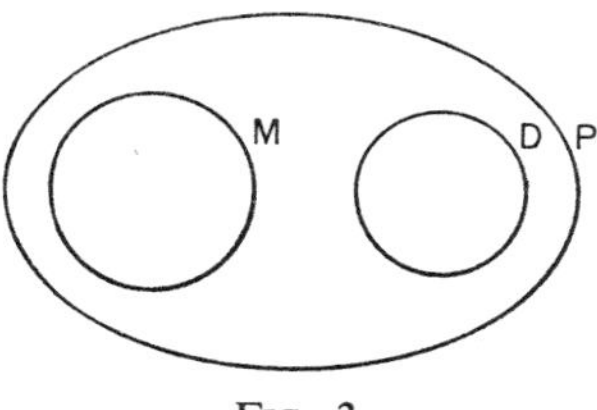

FIG. 3

subsets M and D do not *necessarily* intersect at all. In other words there are not *necessarily* any problems which are both mathematical *and* difficult, i.e. the conclusion does not necessarily follow from the first two statements. We can expose the flaw in the argument by using the laws of sets, for:

$$P \cup M = P \quad \text{and} \quad P \cup D = P$$
$$\therefore \ (P \cup M) \cap (P \cup D) = P \cap P = P$$

But we have shown that:

$$(P \cup M) \cap (P \cup D) = P \cup (M \cap D)$$
$$\therefore \ P \cup (M \cap D) = P$$

but $$P \cup \emptyset = P$$

Hence $M \cap D$ *may* have the value $\emptyset$

i.e. M and D may not necessarily intersect.

Another case which can arise is that in which a true conclusion is reached by a valid argument from two false premises. We say that "two wrongs do not make a right", but it can happen in mathematics. For from the obviously untrue statements $3 > 5$ and $5 = 2$, we obtain the correct result $3 > 2$.

As a final example let us look at the argument:

All children are happy.
Happy people are never mean.
Therefore no children are mean.

If H is the set of happy people, C the set of children and M the set of mean people, we see from Fig. 4 that the argument is valid,

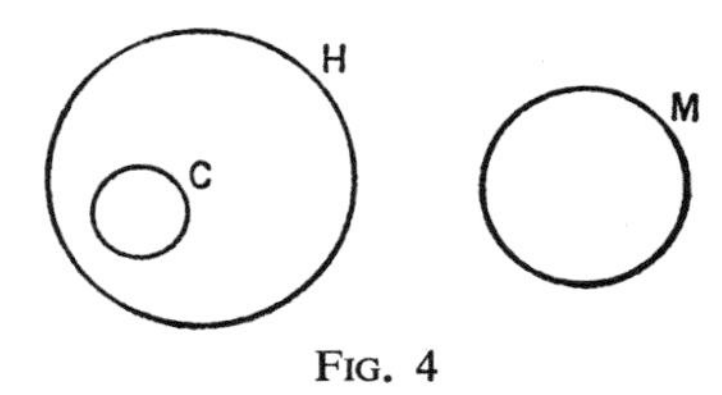

FIG. 4

and the whole statement has truth value T. Alternatively, without using a diagram, we may write:

$$C \cap H = C \quad \text{and} \quad H \cap M = \emptyset$$

Now
$$\begin{aligned} C \cap M &= (C \cap H) \cap M \\ &= C \cap (H \cap M) \text{ (as we saw in Chapter 1)} \\ &= C \cap \emptyset \\ &= \emptyset \end{aligned}$$

Therefore no people are *both* mean and children, i.e. no children are mean.

Exercise 4(a)

In the following arguments, if the conclusion *necessarily* follows from the premises, write T (true); if not, write F (false). Draw Venn diagrams in each case to check your answer. See if you can establish whether or not the arguments are valid by using the laws of sets.

1. Some polygons are rectangles.
Some polygons are squares.
Therefore some rectangles are squares.

2. Some polygons are parallelograms.
Some parallelograms are rhombuses.
Therefore some polygons are rhombuses.

3. Some pentagons have equal angles.
Some pentagons have equal sides.
Therefore some pentagons have equal sides and equal angles.

4. Some cyclic quadrilaterals (C) are trapezia (T).
All rectangles (R) are cyclic quadrilaterals.
Therefore some trapezia are rectangles.

Is this conclusion (a) valid, (b) true? Draw the Venn diagram according to the argument and then redraw it as it should be drawn according to any additional knowledge of geometry you possess. In the second diagram, name the figures which are contained in the sets (c) $R \cap T$, (d) $R \cap C$, (e) $C \cap T$.

5. Some cyclic quadrilaterals (C) are trapezia (T).
Some rhombuses (R) are cyclic quadrilaterals.
Therefore some trapezia are rhombuses.

Is this conclusion (a) valid, (b) true? Draw the Venn diagram showing the sets C, T, R, as they should appear. Name the figures which are contained in the sets (c) $R \cap T$, (d) $R \cap C$, (e) $R \cap T \cap C$.

6. Some boys are tall.
Some boys have red hair.
Therefore some red haired boys are tall.

7. Some gems are expensive.
All diamonds are expensive.
Therefore some gems are diamonds.

8. All sculptors are artists.
Some Frenchmen are sculptors.
Therefore some Frenchmen are artists.

9. All sculptors are artists.
Some Frenchmen are sculptors.
Therefore some French artists are not sculptors.

10. Some rare gases are inert.
Argon is a rare gas.
Therefore Argon is an inert gas.

11. Some footballers can play tennis.
Some tennis players can play cricket.
Therefore some footballers can play cricket.

12. Paris is in Yorkshire.
Yorkshire is in England.
Therefore Paris is in England.

13. Madrid is in Surrey.
Surrey is in Spain.
Therefore Madrid is in Spain.

14. Some teachers teach mathematics or physics or chemistry.
Some mathematics teachers can teach physics.
Some physics teachers can teach chemistry.
Therefore some mathematics teachers can teach chemistry.

15. Some animals are large.
Some animals are wild.
Therefore some wild animals are large.

16. Animals are either wild or tame.
Animals are either large or small.
Therefore some tame animals are large.

17. Animals are either wild or tame.
Animals are either large or small.
Some tame animals are large.
Therefore some wild animals are small.

18. Mathematical problems are either algebraic or geometrical.
Mathematical problems are either easy or difficult.
Some geometrical problems are easy and some are difficult.
Therefore some algebraic problems are difficult.

19. $x < 1$
$y < 1{,}000$
$\therefore\ x < y$

20. $x < 2$
$x > 7$
$\therefore\ 7 < 2$

21. $y < x$
$y^2 > x^2$
$\therefore\ x < 0$

22. All cats are bats.
Bats wear hats.
Therefore all cats wear hats.

In each of the following examples state a valid conclusion which may be drawn from the given premises.

23. No intelligent people are bad tempered.
Some intelligent people are red haired.

24. Some old things are valuable.
Valuable things are beautiful.

25. In our village all the white cows are branded J.
Cows branded J belong to Farmer Jones.
Only black cows wear bells.

26. All sensible people wear good shoes.
People who wear bad shoes do not walk a lot.
All people who walk a lot are healthy.

27. Numbers which are not integers cannot be natural numbers.
Some rational numbers are integers.
Rational numbers are special types of real numbers.
Some complex numbers are real.

SENTENCE LOGIC

The two arguments:

Jack is a Yorkshireman.
All Yorkshiremen are Englishmen.
Therefore Jack is an Englishman.

and:

Saltpetre is a nitrate.
All nitrates are compounds.
Therefore saltpetre is a compound.

as we see from Fig. 5 and Fig. 6 have precisely the same *structure.* In the one case we may write:

$$J \in Y \subset E$$

and in the other:

$$S \in N \subset C.$$

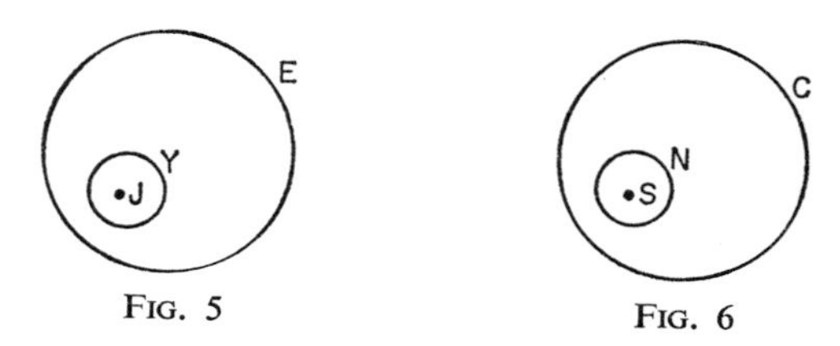

FIG. 5 FIG. 6

From a mathematical point of view we are not really interested in Jack or in saltpetre, but we are interested in the sort of argument involving them. We could put

$$J = S = p, \qquad Y = N = Q, \qquad E = C = R,$$

and say that in both cases the argument has the form

$$p \in Q \subset R.$$

In this section we shall condense the arguments by representing a statement by a small letter. Thus:

Jack is a Yorkshireman.	Statement "p"
All Yorkshiremen are Englishmen . . .	Statement "q"
Jack is an Englishman	Statement "r"

may be condensed to:

IF p *and* q THEN r.

We now take a further step and introduce new notation for words such as "*and*" and "*if . . . then*". Suppose we have two statements p and q, then

(i) The *negation* of p is written as $\sim p$ (i.e. *not* p).
e.g. if p stands for "Jack is ill", then $\sim p$ stands for "Jack is not ill".

(ii) To represent p *and* q we write $p \wedge q$.
e.g. If p stands for "Mary likes ice-cream" and q stands for "Mary likes sweets", then $p \wedge q$ stands for "Mary likes ice-cream *and* she likes sweets".

(iii) To represent p *or* q (*or both*) we write $p \vee q$.
e.g. If p stands for "The baby is crying" and q stands for "The baby is kicking", then $p \vee q$ stands for "The baby is either crying or kicking or doing both together".

There is a complete analogy between sets and statements, and in fact the laws by which we manipulate statements are the laws of sets. To remind us of this the notation is similar.
e.g. If P is the set of all people who like ice-cream and Q is the set of all people who like sweets, then $P \cap Q$ is the set of all people who like ice-cream and sweets.
In the example (ii) above, we are interested in only one member of these sets, namely Mary. She is in both, i.e. in $P \cap Q$, and so we write $p \wedge q$ to express her taste in luxuries.

We may connect statements by the symbols of *implication* and *equivalence*.

(iv) If the statement p *implies* the statement q, we write $p \rightarrow q$.
e.g. If p stands for "Sam is a New Yorker" and q stands for "Sam is an American", then $p \rightarrow q$, for the first statement *implies* the second. But we cannot write $q \rightarrow p$; not all Americans are New Yorkers!

(v) If, however, we have two statements p and q such that p *implies* q *and* q *implies* p, then we say that the statements are *equivalent* and we write $p \leftrightarrow q$.
e.g. If p stands for "Triangle ABC has all three angles equal" and q stands for "Triangle ABC has all three sides equal", then $p \leftrightarrow q$.

Finally, two statements taken together may imply or be equivalent to a third.

e.g. (a) Jack is a Yorkshireman. . . . . p
All Yorkshiremen are Englishmen . . . q
Jack is an Englishman . . . . r

These statements may be connected by writing $p \wedge q \rightarrow r$.

Suppose we have the statements:

Switch p is closed . . . . p
Switch q is closed . . . . q
Current flows in the circuit . . . c

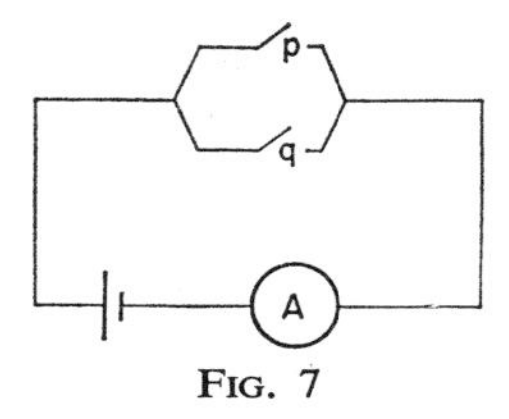

FIG. 7

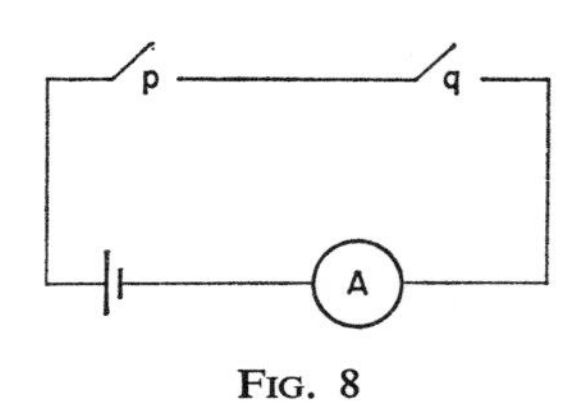

FIG. 8

Then in Fig. 7 we may say, "If either p or q or both are closed, then current flows"; also, "If current flows, then either p or q or both switches are closed", i.e.

$$p \vee q \leftrightarrow c$$

In Fig. 8, however, the full sentence becomes "If both p and q are closed, then current flows"; also, "If current flows, then both p and q must be closed", i.e.

$$p \wedge q \leftrightarrow c$$

Exercise 4(*b*)

1. If a represents the statement "I like tennis" and b represents the statement "I am a good cricketer", write out the meaning of the following sentences:

(*a*) $a \wedge b$ (b) $a \vee b$ (c) $a \wedge \sim b$
(d) $\sim a \vee b$ (e) $\sim a \wedge \sim b$ (f) $a \rightarrow b$

2. If p represents the statement "History is bunk" and q represents the statement "Religion is the opium of the people", write in full the meaning of the sentences:

(a) $p \wedge q$ (b) $p \vee q$ (c) $\sim p \wedge q$
(d) $p \vee \sim q$ (e) $\sim p \vee q$ (f) $p \wedge \sim q$
(g) $\sim p \wedge \sim q$ (h) $\sim p \vee \sim q$ (i) $p \leftrightarrow q$

With which of these sentences do you agree? (You need not put your answer in writing!)

3. If p stands for "I am wearing a green hat" and q stands for "I am wearing black shoes", write in full the meaning of the following sentences:

(a) $\sim p$ (b) $p \vee \sim p$ (c) $q \wedge \sim q$
(d) $p \vee \sim q$ (e) $p \wedge \sim q$ (f) $p \rightarrow q$
(g) $\sim p \wedge q$ (h) $(p \wedge \sim q) \vee (\sim p \wedge q)$

Which of these sentences is nonsensical?

4. Suppose we have the following statements:

Triangles ABC, XYZ are congruent . . . . p
Triangles ABC, XYZ are isosceles . . . . . q
Triangles ABC, XYZ are right-angled . . . r
Triangles ABC, XYZ are similar . . . . . s

Write in full the meaning of the following sentences:

(a) $p \wedge q$ (b) $q \wedge r$ (c) $p \wedge q \wedge r$
(d) $q \vee s$ (e) $r \vee s$ (f) $p \vee s$
(g) $p \rightarrow s$ (h) $r \rightarrow q$ (i) $r \wedge q \rightarrow s$

Which of these sentences are incorrect (or logically false)?

5. Suppose we have the following statements:

ABCD is a parallelogram . . . . a
One pair of the adjacent sides of ABCD are equal in length . . . . b
The diagonals of ABCD are equal in length . c
ABCD is a rhombus . . . . p
ABCD is a rectangle . . . . q
ABCD is a cyclic quadrilateral . . . . . . r
ABCD is a square . . . . s

Write in full the meaning of the following statements:

(a) $p \to a$ (b) $q \to r$ (c) $q \leftrightarrow r$
(d) $a \wedge b \leftrightarrow p$ (e) $a \wedge c \leftrightarrow q$ (f) $c \to q$
(g) $p \vee q \to a$ (h) $p \wedge q \to b \wedge c$ (i) $a \wedge \sim b \to \sim p$
(j) $r \to \sim a$ (k) $p \wedge q \leftrightarrow s$ (l) $a \wedge b \wedge c \leftrightarrow s$
(m) $s \to r$ (n) $\sim s \to \sim p$ (p) $\sim q \leftrightarrow \sim s$

Which of these statements are true (T) and which are false (F)?

6.

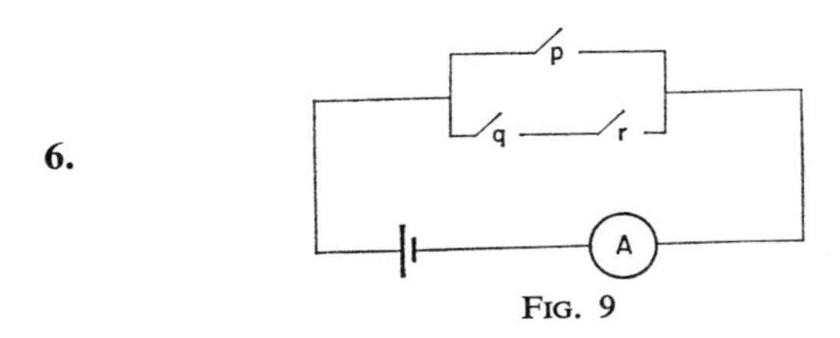

FIG. 9

With reference to Fig. 9, let us suppose that we have the following statements:

Switch p is closed . . . . . . p
Switch q is closed . . . . q
Switch r is closed . . . . r
Current flows in the circuit . . . c

Which of the following statements is logically true:

(a) $p \to c$ (b) $q \to c$ (c) $r \to c$
(d) $q \wedge r \to c$ (e) $q \vee r \to c$ (f) $p \vee q \to c$
(g) $q \wedge r \leftrightarrow c$ (h) $p \vee (q \wedge r) \leftrightarrow c$

TRUTH TABLES

In the previous examples you will probably have had little difficulty in deciding whether or not the sentences were true. In the first place the sentences referred to simple geometrical situations which were familiar to you, and secondly the sentences themselves were not complicated. Situations do arise, however, in which it is not always possible to decide whether a sentence is true or not. In such cases it is helpful to build up a *truth table*.

(i) The simplest assertion is that of negation. Obviously, if a statement p is true (T), then $\sim p$ must be false (F) and vice-versa. The table showing the possible cases is given below. It is the simplest form of truth table.

p	$\sim p$
T	F
F	T

(ii) In the case $p \wedge q$ four cases arise. p and q are either (a) both true, (b) both false, (c) p true and q false, (d) p false and q true. Clearly, "*p and q*" is only a true sentence if p and q are both true. Hence we have a truth table as shown below:

p	q	$p \wedge q$
T	T	T
T	F	F
F	T	F
F	F	F

(iii) The case $p \vee q$ is rather more complicated. Here we are stating that either p or q, or both, are true. Clearly, if p is false and q is true, or vice-versa, the statement or *disjunction* is true.

e.g. (a) "Either spanners are tools *or* monkeys are fish" is a true statement because, at any rate, one of the facts is true.

If both statements are false, then clearly the whole sentence is false.

(b) "All dogs are mad *or* all men are sane" is clearly false.

If, however, both statements are true, a difficulty arises. If I *am* wearing brown shoes and I *am* wearing a green hat, is it true to say "I am wearing brown shoes *or* a green hat"? Now we have defined $p \vee q$ as "*p* or *q or both*". This is called the *inclusive disjunction.* We can in fact say, "I am wearing brown shoes *or* a green hat *or both*", and if both statements are true then the whole compound statement or sentence is also true.

If, however, p stands for "I am in London" and q stands for "I am in Bradford", then the inclusive disjunction has no meaning, for obviously I cannot be either in London or Bradford or in *both* places at the same time. For cases like this mathematicians use the symbol $\veebar$, where $p \veebar q$ means p or q *but not both*. This is called the *exclusive disjunction*. In this chapter we shall use only the inclusive disjunction. The truth table now becomes:

p	q	$p \vee q$
T	T	T
T	F	T
F	T	T
F	F	F

Examples. (*a*) Construct a truth table for $(p \wedge \sim q) \vee (\sim p \wedge q)$.

p	q	$\sim p$	$\sim q$	$p \wedge \sim q$	$\sim p \wedge q$	$(p \wedge \sim q) \vee (\sim p \wedge q)$
T	T	F	F	F	F	F
T	F	F	T	T	F	T
F	T	T	F	F	T	T
F	F	T	T	F	F	F

(*b*) Construct a truth table for $(p \vee q) \vee \sim p$.

p	q	$\sim p$	$p \vee q$	$(p \vee q) \vee \sim p$
T	T	F	T	T
T	F	F	T	T
F	T	T	T	T
F	F	T	F	T

In this case $(p \vee q) \vee \sim p$ is a sentence which is true under all possible circumstances. We say that it is *logically true*. If p stands for "I am wearing brown shoes" and q stands for "I am wearing a green hat", then $(p \vee q) \vee \sim p$ reads "Either I am wearing brown shoes or a green hat or both, or, I am not wearing brown shoes", which, when you think about it, is true no matter what I am wearing.

(iv) In sentence logic the symbol $\rightarrow$ stands for "if . . . then". The most interesting cases arise when q is a logical consequence of p and $p \rightarrow q$ has the fuller meaning "p implies q". We can, however, make sentences using the words "if . . . then" without the fullest meaning of implication being involved. For example if p stands for "It will rain on Saturday" and q for "I shall be annoyed", then $p \rightarrow q$ reads "If it rains on Saturday then I shall be annoyed". Let us decide that if all this happens we have *p true, q true, then $p \to q$ is true.* If q is false the sentence reads "If it rains on Saturday then I shall not be annoyed", and this is a contradiction. *Hence if p is true and q is false, then $p \to q$ is false.* It is not very clear, however, what happens if p is false. If q is true the sentence now reads "If it does not rain on Saturday then I shall be annoyed". This may well be a true sentence (it is not only bad weather that upsets me!), so we resolve the doubt by taking *p false, q true, then $p \to q$ is true.* Finally, what happens if both p and q are false? The sentence now has the meaning "If it does not rain on Saturday then I shall not be annoyed". This has more or less the same sense as the case p true q true, so although it is still rather a doubtful statement, we make up our minds to take *p false, q false, then $p \to q$ is true.*

Summarizing this not very rigorous discussion we have the truth table:

p	q	$p \to q$
T	T	T
T	F	F
F	T	T
F	F	T

(v) The final connective is $p \leftrightarrow q$, or "if p then q, and if q then p". Here there is no doubt at all. p and q can only be *equivalent* statements when both are true or when both are false. Thus:

p	q	$p \leftrightarrow q$
T	T	T
T	F	F
F	T	F
F	F	T

Exercise 4(c)

Construct truth tables for each of the following sentences:

1. $\sim q$ **2.** $p \wedge \sim q$ **3.** $\sim p \wedge q$
4. $\sim p \wedge \sim q$ **5.** $p \vee \sim p$ **6.** $p \vee \sim q$
7. $\sim p \vee q$ **8.** $(p \wedge q) \vee (\sim p \wedge \sim q)$
9. $p \to \sim q$ **10.** $\sim p \leftrightarrow \sim q$

11. By constructing their truth tables show that the sentences $p \to q$ and $\sim p \vee q$ are equivalent.

12. Show that $p \to q$ and $\sim q \to \sim p$ have the same truth set.

13. By constructing their truth tables show that $(p \wedge \sim q) \to \sim p$ is equivalent to $p \to q$. Make up statements for p and q and show that the sentences $p \to q$ and $(p \wedge \sim q) \to \sim p$ have the same meaning.

14. Find a simpler statement which is equivalent to $(p \wedge \sim q) \to q$.

15. Show that the following sentences are all logically true:

(a) $p \vee \sim p$ (b) $p \vee (\sim p \vee q)$ (c) $(p \leftrightarrow q) \to (p \to q)$

16. Construct a sentence which is logically false.

BOOLEAN ALGEBRA

A sentence or expression such as $(p \wedge q) \vee (\sim p \wedge \sim q)$, although a very simple one, looks quite complicated to unfamiliar eyes. In the notation of sentence logic a complicated sentence can look really formidable. Since we shall be doing rather more manipulation in the remainder of this chapter we shall introduce a rather simpler notation which is often used in the algebra of switching circuits.

Instead of $\sim a$ we shall write a'.
Instead of $a \vee b$ we shall write $a+b$.
Instead of $a \wedge b$ we shall write $a.b$.

A comparison of the symbols used in sets, sentence logic and switching circuits is given in the table below:

	Complement *Negation*	*Union* *"Either — or — or both"*	*Intersection* *"and"*
Sets	A'	$A \cup B$	$A \cap B$
Sentence logic	$\sim a$	$a \vee b$	$a \wedge b$
Switching circuits	a'	$a + b$	$a.b$

Both the algebra of sentence logic and the algebra of switching circuits obey the laws of sets. They are examples of Boolean algebra, so called after a great English mathematician, George Boole (1815–64).

We now make a list of all the laws of sets which we discovered or proved in Chapter 1.

If $\mathscr{E}$ is some universal set with subsets A, B and C we have:

1. $A \cap \mathscr{E} = A$
2. $A \cup \emptyset = A$
3. $A \cap (B \cup C) = (A \cap B) \cup (A \cap C)$
4. $A \cup (B \cap C) = (A \cup B) \cap (A \cup C)$
5. $A \cap (A \cup B) = A$
6. $A \cup (A \cap B) = A$
7. $A \cap \emptyset = \emptyset$
8. $A \cup \mathscr{E} = \mathscr{E}$
9. $A \cap A = A$
10. $A \cup A = A$
11. $(A \cup B)' = A' \cap B'$
12. $(A \cap B)' = A' \cup B'$

If we now rewrite these laws in terms of the new notation, and if in addition we write 1 instead of $\mathscr{E}$ and 0 instead of $\emptyset$, we obtain the rules of Boolean algebra:

1. $a \times 1 = a$
2. $a + 0 = a$
3. $a(b+c) = ab+ac$
4. $a+bc = (a+b)(a+c)$
5. $a(a+b) = a$
6. $a+ab = a$
7. $a \times 0 = 0$
8. $a+1 = 1$
9. $a \times a = a$
10. $a+a = a$
11. $(a+b)' = a'.b'$
12. $(ab)' = a'+b'$

Some of these rules, such as 1, 2, 3 and 7, are clearly reminiscent of ordinary algebra, but the others are quite different. Furthermore, they are not all independent. Rule 3 states that we can remove brackets in the normal way. Applying this to the left-hand side of rule 5, we have a^2+ab. But from rule 9, $a^2 = a$, hence rule 5 becomes $a(a+b) = a^2+ab = a+ab = a(1+b)$. Now by rule 8, $1+b = 1$, hence $a+ab = a$, which is rule 6. Thus rule 6 follows from rules 3, 9 and 8.

Again, the right-hand side of rule 4

$$\begin{aligned} &= a(a+c)+b(a+c) \text{ from } 3 \\ &= a+b(a+c) \text{ from } 6 \\ &= a+ab+bc \text{ from } 3 \\ &= a+bc \text{ from } 6 \end{aligned}$$

and this is rule 4.

A mathematician bent on economy could condense these twelve rules to a smaller number of independent ones. However, they are all true, and in the early stages it is perhaps more helpful to have a wide choice.

Another interesting feature of this algebra is that it incorporates the *principle of duality*. If we take any rule and interchange $\times$ and $+$, 0 and 1, we obtain another rule which is true. Thus rule 1 is the *dual* of rule 2, rule 3 is the *dual* of rule 4, rule 5 is the *dual* of rule 6, rule 7 is the *dual* of rule 8, rule 9 is the *dual* of rule 10 and rule 11 is the *dual* of rule 12. The twelve rules contain six pairs of *self-dual relationships*.

Exercise 4(*d*)

Using the rules of Boolean algebra, simplify the following expressions as much as possible:

1. $a^2(a+b)$ **2.** $a+a^2$ **3.** a^3+b^3

4. a^3+a^2+a+1 **5.** $a^2+2ab+b^2$ **6.** a^2+ab+b^2

7. $a(a+b)(a+c)$ **8.** $a(a+b)(a+c)(a+d)(a+e)(a+f)$

9. $(a+1)^{10}$ **10.** a^2b+ab^2

Just as for sets, $A \cap A' = 0$, $A \cup A' = \mathscr{E}$, so we have $a.a' = 0$ and $a+a' = 1$. Using these results and any of rules 1–12, simplify the following:

11. $(a+b)(a+b')$ **12.** $a'(a+b)$ **13.** $a'b'(a+b)$
14. $a'b'+a+b$ **15.** $(a+b+x'+y')(a+b+xy)$
16. $(a+b+x'+y'+z')(a+b+xyz)$
17. $(a+b'+c')(a+bc)$
18. $(a+b+c+d')(a+b+c+d)(a+b+c')(a+b')$

APPLICATION TO LOGIC

Example. Confronted with the following set of regulations, a motorist, though anxious to conform to the law, was nevertheless understandably perplexed. "When you are not on the left, sound the horn. If you keep to the left and sound the horn, do not stop. If you are stationary, or you are on the right, do not sound the horn".

We can simplify as follows. Denote

"you keep to the left" by l,
"you should sound the horn" by h,
"you should stop" by s.

The statements then reduce to:

$$l' \rightarrow h$$

$$l \wedge h \rightarrow s'$$

$$s \vee l' \rightarrow h'$$

Or, in the new notation,

$$l' = h$$

$$lh = s'$$

$$s+l' = h'$$

or

$$l'h' = 0$$

$$lhs = 0$$

$$h(s+l') = 0$$

Adding together:

$$l'h'+lhs+hs+l'h = 0$$

$$\therefore\ l'(h'+h)+hs(l+1) = 0$$

Now $\qquad h'+h = 1 \quad \text{and} \quad l+1 = 1$

$$\therefore\ l'+hs = 0$$

$$\therefore\ (l'+hs)' = (0)'$$

$$\therefore\ l(hs)' = 1$$

And this reads:

"Always drive on the left and do not sound the horn when stationary".

Exercise 4(*e*)

1. Obtain the main conclusion from the premises:

In our village, all the white cows are branded J.
Cows branded J belong to Farmer Jones.
Only black cows wear bells.

2. Simplify the following set of rules:

Boys may wear no tie and/or no white shirt.
Boys may wear no tie and/or a white shirt.
Boys may wear a tie and/or a white shirt.

3. Five independent witnesses stated respectively that the man seen running away from the scene of the crime was (a) tall, dark and handsome, (b) tall, dark but not handsome, (c) tall and handsome but not dark, (d) dark but not tall, (e) handsome but neither tall nor dark. What sort of man should the police start looking for?

SWITCHING CIRCUITS

Suppose we have the statements:

Switch p is closed . . . . p
Switch q is closed . . . . q
Current flows in the circuit . . . c

Then for Fig. 10 (i) we may write:

$$p \wedge q \leftrightarrow c \quad . \quad . \quad . \quad . \quad . \quad . \quad . \quad . \quad . \quad . \quad . \quad 1$$

and for Fig. 10 (ii) we have:

$$p \vee q \leftrightarrow c \quad . \quad . \quad . \quad . \quad . \quad . \quad . \quad . \quad . \quad . \quad . \quad 2$$

Expressing these statements in our new notation we have:

$$pq = 1 \quad . \quad . \quad . \quad . \quad . \quad . \quad . \quad . \quad . \quad . \quad . \quad . \quad 1$$
$$p+q = 1 \quad . \quad . \quad . \quad . \quad . \quad . \quad . \quad . \quad . \quad . \quad . \quad 2$$

or (1) If p *and* q are closed then current flows and (2) If either p *or* q *or both* are closed, current flows.

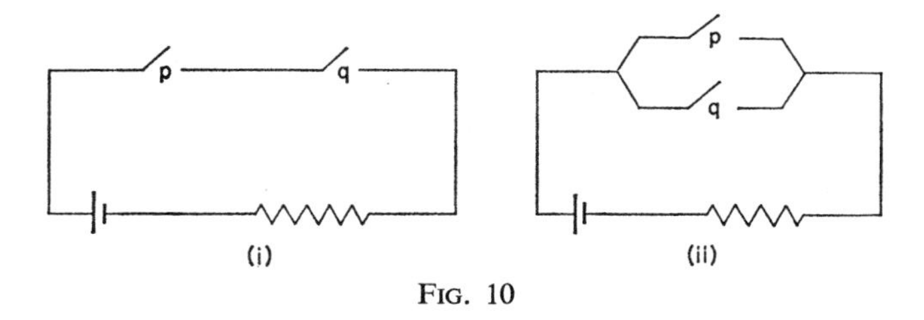

FIG. 10

In switching circuits we can see what happens for the various cases p, q, open or closed, if we construct *a closure table.* Such tables are really truth tables, except that instead of T (true) we write 1 ("switch closed" or "current flows"), and instead of F we write 0 ("switch open" or "no current flows"). Closure tables for the cases (i) and (ii) above are shown below.

p	q	pq	$p+q$
1	1	1	1
1	0	0	1
0	1	0	1
0	0	0	0

These are exactly analogous to the truth tables previously obtained for $p \wedge q$ and $p \vee q$.

In switching Boolean algebra we denote a switch by a single letter $a, b \ldots x, y$. If two switches open and close simultaneously we denote them by the same letter. If one is open when the other is closed and vice-versa, we denote one by x and the other by x'. Switches x and y in *parallel* are denoted by $x+y$. Switches in *series* are denoted by xy. With these rules we can build up a Boolean function for any circuit which involves sets of switches connected either in series or in parallel.

Example (i).

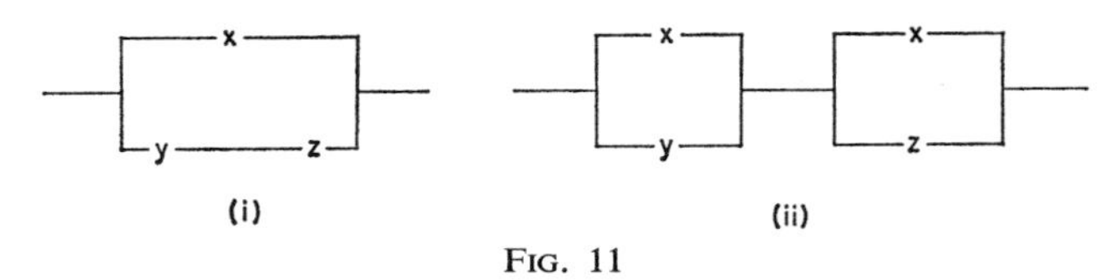

FIG. 11

The Boolean function for circuit (i) is $x+yz$.
The Boolean function for circuit (ii) is $(x+y)(x+z)$.
Closure tables for these two functions are shown below:

x	y	z	$x+y$	$x+z$	yz	$x+yz$	$(x+y)(x+z)$
1	1	1	1	1	1	1	1
1	1	0	1	1	0	1	1
1	0	1	1	1	0	1	1
0	1	1	1	1	1	1	1
1	0	0	1	1	0	1	1
0	1	0	1	0	0	0	0
0	0	1	0	1	0	0	0
0	0	0	0	0	0	0	0

The functions $x+yz$ and $(x+y)(x+z)$ have the same values for all possible values of x, y and z. Hence the functions, and therefore the circuits, are equivalent. We should, of course, expect this, since

$$x+yz = (x+y)(x+z) \text{ is rule 4.}$$

In fact, switching circuits obey all the rules of Boolean algebra, hence we may use any one of these to simplify or design a circuit.

Example (ii). Simply the following circuits:

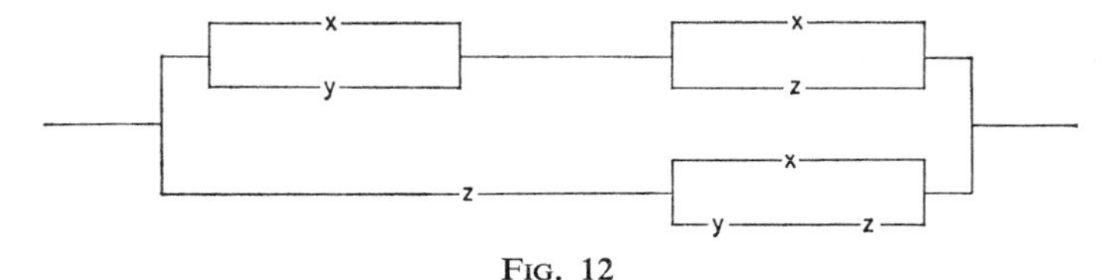

FIG. 12

In Fig. 12 the Boolean function of the circuit is

$$(x+y)(x+z)+z(x+yz)$$

or $\quad x+yz+z(x+yz) \quad$ by rule 4

or $\quad (x+yz)(1+z) \quad$ by rule 3

or $\quad x+yz \quad$ by rule 8

Hence the circuit in Fig. 12 is equivalent to the simpler circuit in Fig. 11 (i).

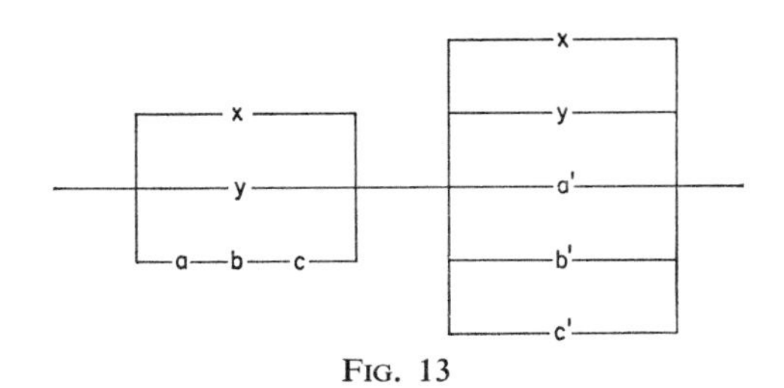

FIG. 13

In Fig. 13 the Boolean function of the circuit is

$$(x+y+abc)(x+y+a'+b'+c')$$

or $\quad (x+y+abc)[x+y+(abc)'] \quad$ by rule 12

or $\quad x+y+(abc)(abc)' \quad$ by rule 4

but $\quad (abc)(abc)' = 0.$

Hence, the circuit in Fig. 13 is equivalent to two switches x, y in parallel.

Example (iii). Design a circuit so that a strip lighting filament in a corridor may be controlled independently by either of two switches placed one at each end of the corridor.

In order to find the required circuit we first construct a closure table. This is shown below.

p	q	*Boolean function*
1	1	1
1	0	0
0	1	0
0	0	1

From the table you will see that we have arranged that the light is on when p and q are both closed, i.e. have value 1. If either p or q changes from 1 to 0, the light must go off. For any further change in either p or q the light must go on again. Therefore current must flow when either p and q are both on, or when p and q are both off, i.e. when either $p \wedge q$, or $\sim p \wedge \sim q$, i.e. when $(p \wedge q) \vee (\sim p \wedge \sim q)$, which in the new notation gives

$$pq+p'q'.$$

This is the Boolean function of the circuit required. The circuit appears as shown in Fig. 14.

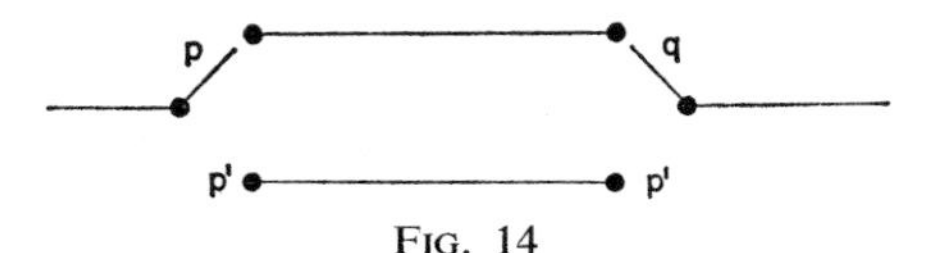

FIG. 14

Exercise 4(f)

Write down the Boolean function for each of the following circuits. Simplify the expressions where possible and sketch in each case the simplified circuit.

1.

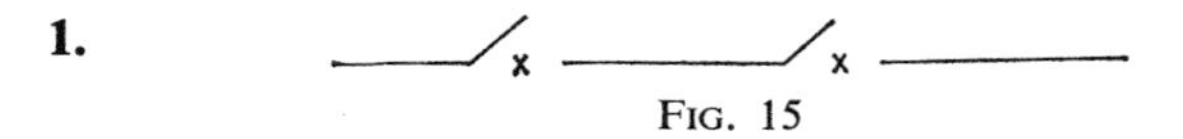

FIG. 15

2.

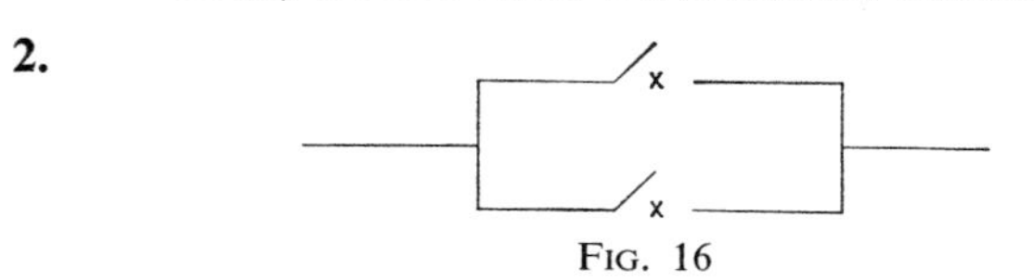

FIG. 16

3.

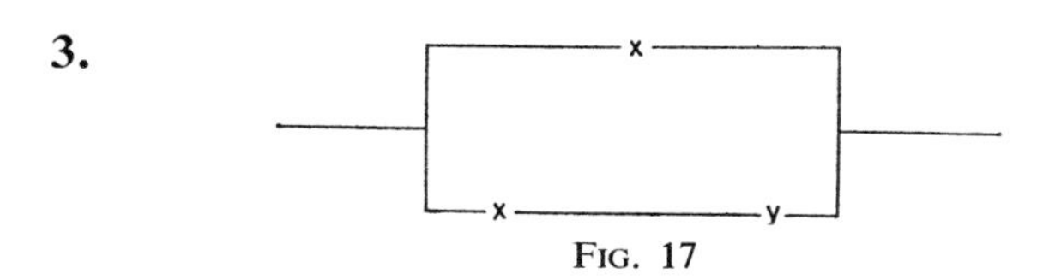

FIG. 17

4.

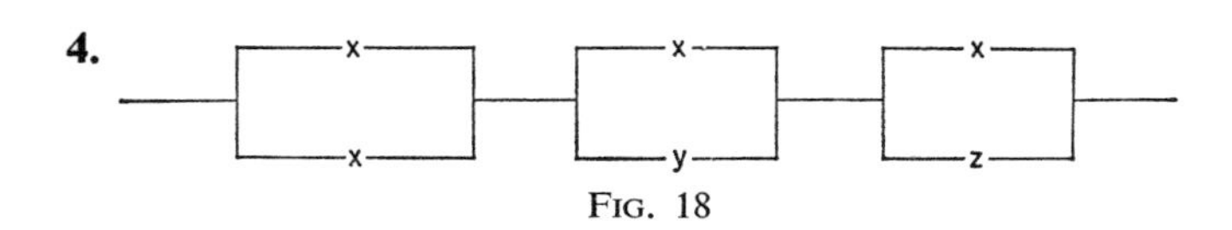

FIG. 18

5.

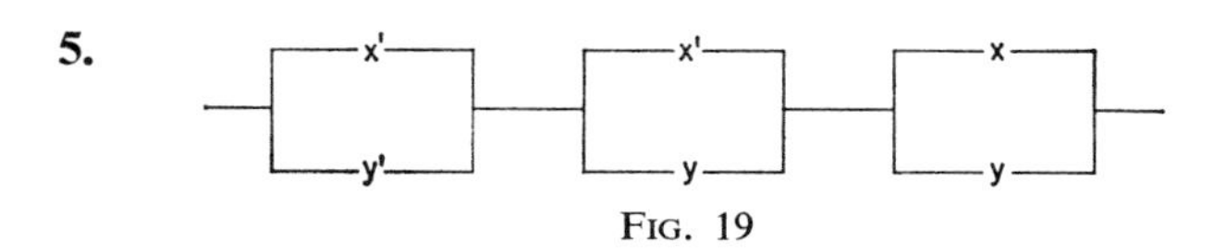

FIG. 19

6.

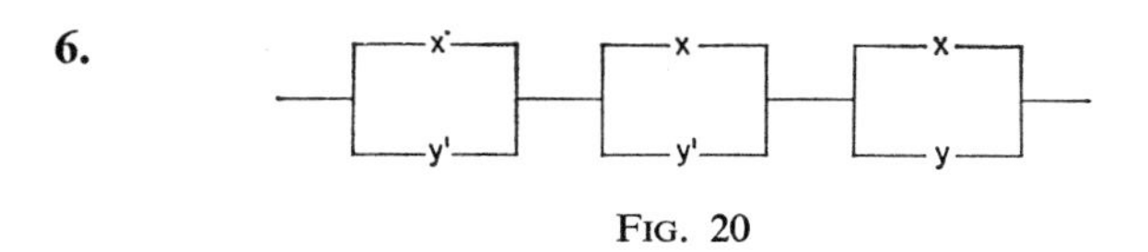

FIG. 20

7.

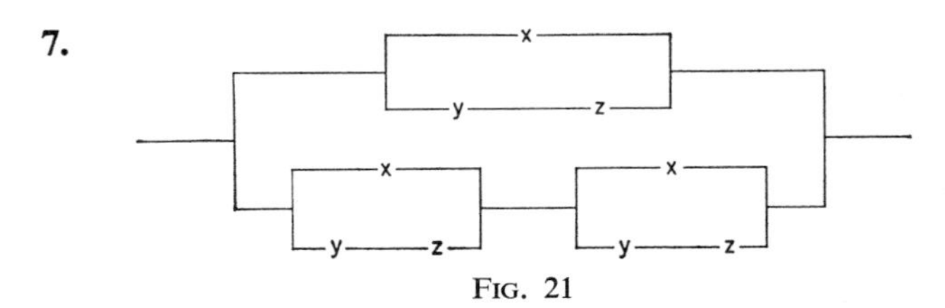

FIG. 21

8.

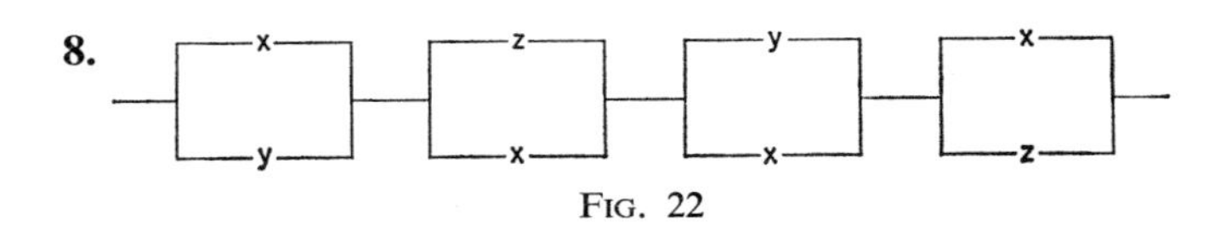

FIG. 22

9.

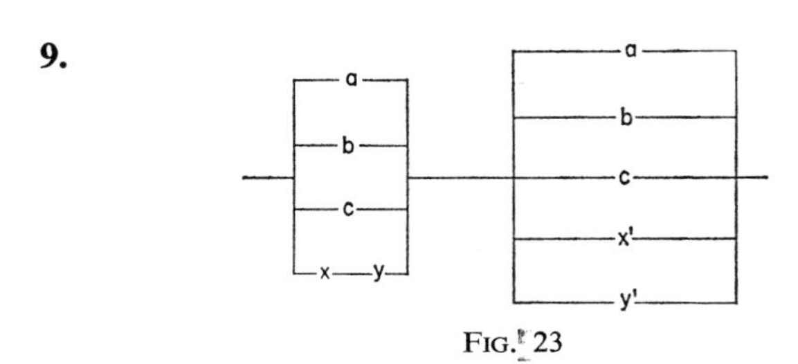

FIG. 23

10.

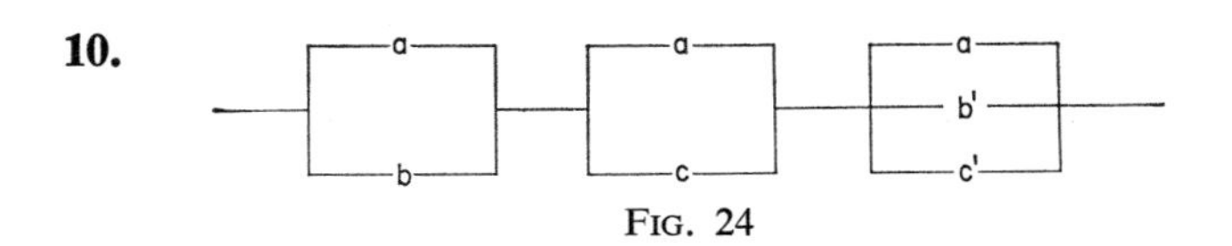

FIG. 24

11.

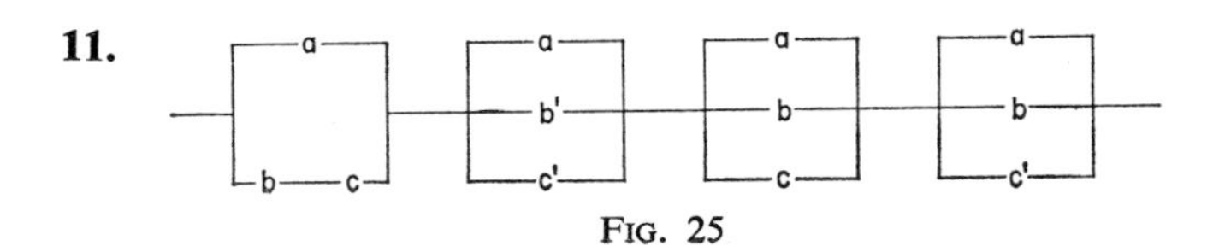

FIG. 25

12.

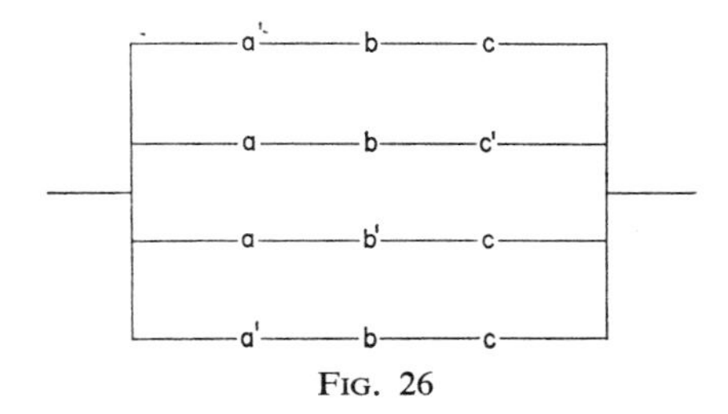

FIG. 26

To what use might this circuit be put? Construct a closure table and examine how the Boolean function changes with successive changes in *a*, *b* and *c*.

13.

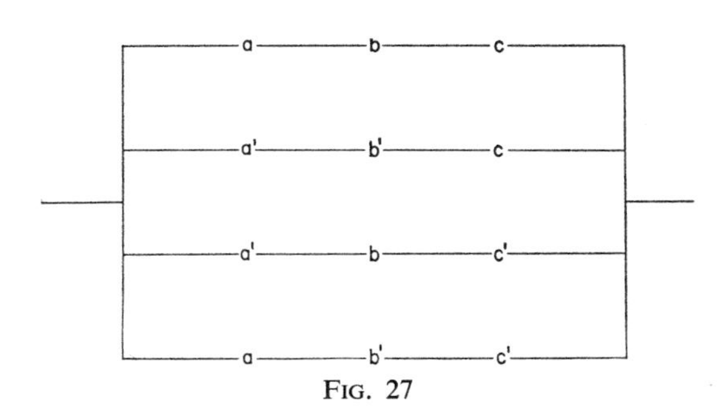

FIG. 27

To what use might this circuit be put? Construct a closure table and examine how the Boolean function changes with successive changes in *a*, *b* and *c*.

14.

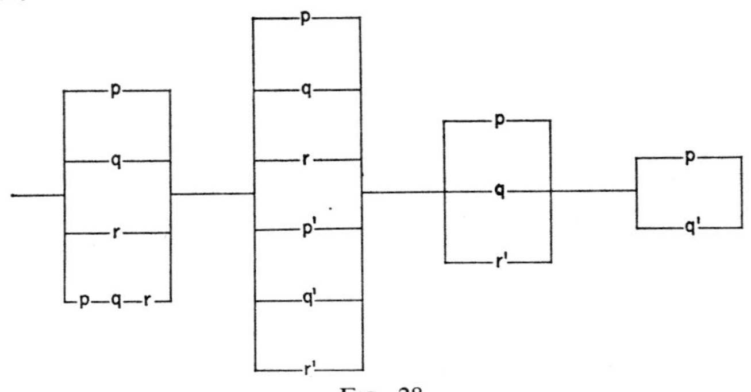

FIG. 28

15.

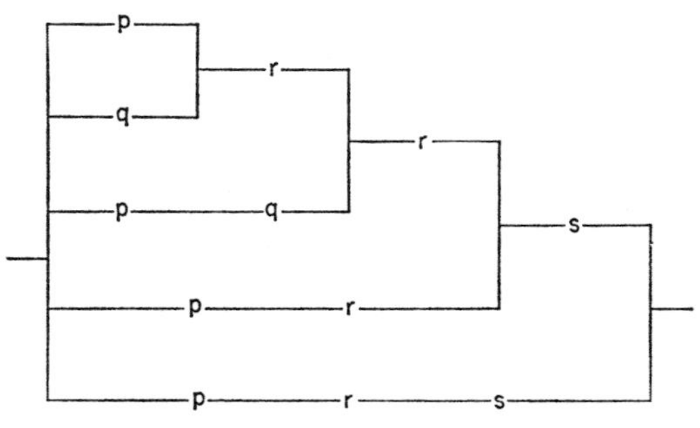

FIG. 29

16.

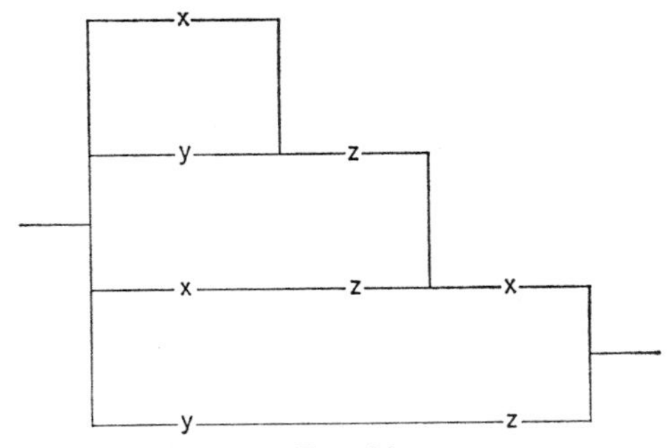

FIG. 30

17.

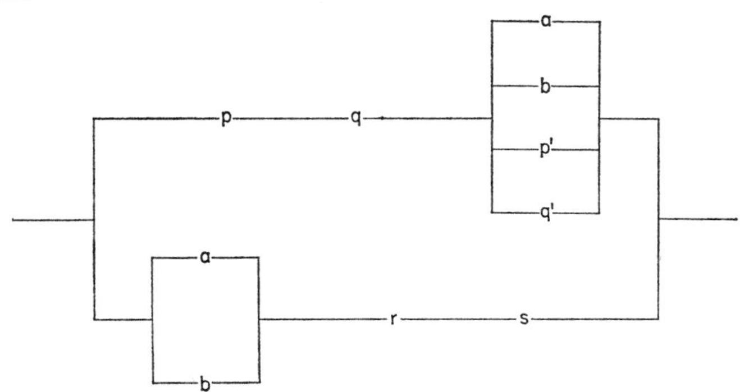

FIG. 31

18.

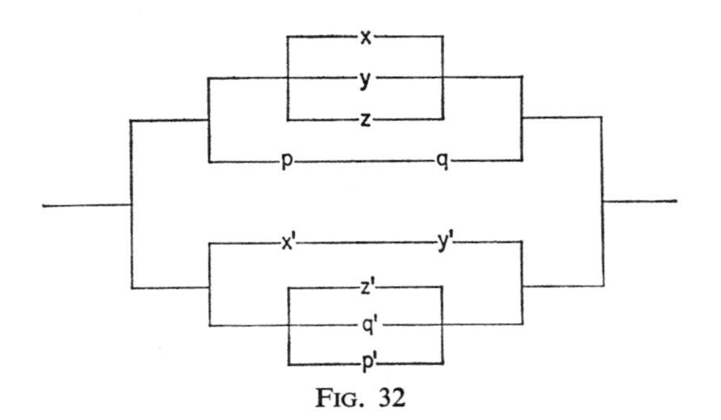

FIG. 32

19. A corridor is illuminated by a long strip lighting element. Design a circuit whereby the light is independently controlled by any one of three switches placed respectively at the two ends and in the middle of the corridor.

20. A panel of three members vote by pressing buttons placed in front of them. Design a circuit in which a buzzer sounds whenever a majority vote occurs.

21. A machine contains three fuses p, q, r. It is desired to arrange them so that if p blows the machine stops, but if p does not blow then the machine only stops when both q and r have blown. Design the required fuse circuit.

22. A committee of five vote by pressing switch buttons a, b, c, d, e. The chairman, who controls switch a, does not vote unless required to give a casting vote, otherwise the decision is by a majority vote of the other four members. Design a switching circuit by means of which a buzzer sounds whenever a resolution is carried.

23. The representatives of five countries, A, B, C, D, E, vote by pressing five buttons a, b, c, d, e. By the constitution, A and C each have a veto, but provided they both vote for a resolution this may be carried by a simple majority of the members. Design the required circuit.

24. A machine is controlled by four fuses a, b, c, d. If three or four fuses blow, the machine stops. If either a and c blow together, or if b and d blow together, the machine stops; otherwise the machine continues to operate. Design the required fuse circuit.

5
SCALES OF NOTATION

THE BINARY SCALE

For normal purposes the great majority of numbers which we use in everyday calculations are based on a scale of ten. Indeed there are very good reasons for adopting a decimal system of measurements for quantities such as weight, length, and money as well. In the first place, such a change would greatly simplify commercial transactions between this country and other countries of the world; secondly, by replacing a hotch potch of tables based on antiquated units by one simple decimal scale of units, we should save an enormous amount of time and effort in our schools.

In the scale of ten, or *denary scale*, we have only one rule for "carrying"; when the digits in any column total ten we carry 1 to the next column. In the denary scale the number 245 *means* 2 hundreds plus 4 tens plus 5 units, or $245 = 2\times10^2+4\times10+5\times1$. The idea of using 10 as a base is intimately connected with the fact that we possess 10 fingers (counting thumbs as fingers!). For some purposes, however, this system has its disadvantages. In order to record any number we require no less than ten different symbols.

We now consider a number system which has only two symbols, 0 and 1. The rule of carrying is that when the digits of any column total two, we carry 1 to the next column. Thus, in this system, 2 becomes 10. Instead of having columns of units, tens, hundreds, etc., we now have columns of units, twos, fours, etc. Place values are ordered not in powers of ten but in powers of two. This scale is called the *binary* scale. In this scale the number 111 *means*

$1 \times 2^2 + 1 \times 2 + 1 \times 1$. That is to say, the number 111 (base 2) = 7 (base 10).

Examples (i). 1011 (base 2) $= 1 \times 2^3 + 0 \times 2^2 + 1 \times 2 + 1 \times 1$
$= 11$ (base 10)

(ii). 11111 (base 2)
$= 1 \times 2^4 + 1 \times 2^3 + 1 \times 2^2 + 1 \times 2 + 1 \times 1$
$= 31$ (base 10)

CONVERTING DENARY NUMBERS TO BINARY NUMBERS

If a greengrocer requires a set of weights capable of weighing any whole number of pounds up to, say, 31 lb, a little reflection will soon convince us that he does not require 31 different weights. In fact all he does need is the following set: {16 lb, 8 lb, 4 lb, 2 lb, 1 lb}. With these five weights he can weigh any quantity from 1 lb to 31 lb. The ways in which he does this are set out in the table on page 79.

What we have done in this table, perhaps unwittingly, is to convert the denary numbers 1 to 31 into binary numbers. We may, of course, continue the process indefinitely. Thus, to express any denary number in the binary scale, we simply express it as the sum of powers of two, e.g.

730 (base 10) $= 512 + 128 + 64 + 16 + 8 + 2$
$= 2^9 + 2^7 + 2^6 + 2^4 + 2^3 + 2^1$
$= 1011011010$ (base 2)

Quantity to be weighed	*Weights required*				
	16 lb 2^4	8 lb 2^3	4 lb 2^2	2 lb 2	1 lb 1
1					1
2				1	0
3				1	1
4			1	0	0
5			1	0	1
6			1	1	0
7			1	1	1
8		1	0	0	0
9		1	0	0	1
10		1	0	1	0
11		1	0	1	1
12		1	1	0	0
13		1	1	0	1
14		1	1	1	0
15		1	1	1	1
16	1	0	0	0	0
17	1	0	0	0	1
18	1	0	0	1	0
19	1	0	0	1	1
20	1	0	1	0	0
21	1	0	1	0	1
22	1	0	1	1	0
23	1	0	1	1	1
24	1	1	0	0	0
25	1	1	0	0	1
26	1	1	0	1	0
27	1	1	0	1	1
28	1	1	1	0	0
29	1	1	1	0	1
30	1	1	1	1	0
31	1	1	1	1	1

Exercise 5(*a*)

1. Convert the following binary numbers into their equivalents in the denary scale:

(a) 10	(b) 11	(c) 101
(d) 1101	(e) 1010	(f) 11010
(g) 10101	(h) 11110	(i) 111101
(j) 101011	(k) 1001101	(l) 110010101
(m) 110101011	(n) 111111	(o) 1000001
(p) 111001010	(q) 111000110010	(r) 1010101010101

2. Convert the following denary numbers into their equivalents in the binary scale:

(a) 3	(b) 7	(c) 11
(d) 29	(e) 39	(f) 56
(g) 63	(h) 126	(i) 132
(j) 111	(k) 255	(l) 513
(m) 608	(n) 768	(o) 777
(p) 1024	(q) 10,000	(r) 1,000,000

A much quicker way of converting denary numbers into their binary equivalents is to divide the given denary number repeatedly by 2 until a quotient of 0 is reached. The binary number required is then the remainders placed in reverse order to that in which they were obtained.

Example. Express 730 (base 10) in the binary scale.

2	730		
2	365	remainder	0
2	182	,,	1
2	91	,,	0
2	45	,,	1
2	22	,,	1
2	11	,,	0
2	5	,,	1
2	2	,,	1
2	1	,,	0
	0	,,	1

(An upward arrow beside the remainders shows they are read from the bottom up.)

Hence 730 (base 10)
= 1011011010 (base 2).

Exercise 5(*b*)

1. Explain how the method of repeated division by 2 works.

2. Repeat Ex. 5(a), Question 2 by the method of repeated division by 2.

ADDITION AND SUBTRACTION OF BINARY NUMBERS

In addition, if the sum of the digits in any column totals 2, we carry 1 to the next column. If this sum totals 4 we can either carry 2 to the next column, or we can carry 1 to the next column but one. If the sum totals 6 we can either carry 3 to the next column, or we can carry 1 to the next column and 1 to the next column but one. It is better to adopt this latter process; in the first place, if we carry 3 this will only have to be converted in the next column, and secondly, in binary arithmetic *there is no such digit as* 3, or, for that matter, 2 either.

Examples.

```
 101101+
  10111
 -------
 1000100
```

This is the equivalent of adding 45 and 23 to obtain 68 in the denary scale.

(ii)

```
    1101+
    1011
   10110
   11111
    1111
 -------
   1111     carrying figures
  1

 1011100   Answer
 -------
```

In the denary scale, 13+11+22+31+15 = 92.

In subtraction we sometimes need to "borrow" from the next column above. When we borrow from this column we are really borrowing "2" if we think in the denary scale, but it is better to keep to the notation of the binary scale and to record the 1 borrowed as "10" in the column next below.

Example.

$$\begin{array}{r} \overset{1}{\not{10}}\;10\;\;10\;\overset{1}{\not{10}}\,\overset{1}{\not{10}}\,10 \\ \not{1}0\not{1}\not{1}000- \\ 111111 \\ \hline 11001 \end{array}$$

This is the equivalent of $88-63 = 25$ in the denary scale.

Exercise 5(*c*)

Perform the following additions in the binary scale. Check each result by converting to the denary scale and doing the additions in the scale of ten.

1. $101+11$

2. $111+101$

3. $1101+1001$

4. $1010+1111$

5. $10101+1010$

6. $11011+10111$

7. $110101101+101011101$

8. $1110101101011+1101000110111$

9. $1101+1011+1110$

10. $1001+1101+1010$

11. $1011+1111+1101+1001$

12. $11101+10111+10001+11011$

13. $1101111+1100011+1001111+1110011$

14. $1+11+111+1111+11111+111111$

15. $1111+1111$

16. $1111+1111+1111+1111$

Perform the following subtractions in the binary scale. Check each result by converting to the denary scale and working in the scale of ten.

17. 101 − 11

18. 1001 − 101

19. 10000 − 1001

20. 10101 − 1010

21. 10000 − 111

22. 1011001 − 101001

23. 1100110 − 111111

24. 10101100110101 − 10010011001001

25. 11110 − 1111 − 1111

26. 1111 − 101 − 101 − 101

27. 100001 − 1011 − 1011 − 1011

28. 1000000000 − 11111111

MULTIPLICATION AND DIVISION OF BINARY NUMBERS

In Question 15 of the last exercise we saw that:

$$1111+1111 = 11110$$

In the denary scale this would be written as:

$$15+15 = 30$$

Repeated addition is more concisely expressed in the form of multiplication, so we may write:

$$15\times 2 = 30$$

Reconverting to the binary scale we have:

$$1111\times 10 = 11110$$

i.e. the rule for multiplying by 10 in the denary scale applies equally well in the binary scale.

Again, in Question 16:

$$\begin{aligned} 1111+1111+1111+1111 &= 1111\times 100 \\ &= 111100 \end{aligned}$$

The multiplication 1111 × 111 is therefore set out and performed in exactly the same way as we should do it when working in the denary scale:

```
   1111
  ×111
-------
   1111
  11110
 111100
-------
1101001
```

The process of division is simply a more compact method of carrying out repeated subtraction. Thus, to find out how many times the binary number 1011 is contained in 100001, we may either perform three (11 binary) repeated subtractions:

```
100001 –
  1011
------
 10110 –
  1011
------
  1011 –
  1011
------
  ....
```

or we may perform a long division in exactly the same way as we should do it with denary numbers:

```
          11
      -------
1011 | 100001
     |  1011
     |  -----
     |   1011
     |   1011
     |   ----
     |   ....
```

Exercise 5(*d*)

Carry out the following multiplications and divisions in the binary scale. Check each result by working the calculation in the denary scale.

1. 110×10

2. 1001×100

3. 1111×1000

4. 111×11

5. 1011×1010

6. 11111×1110

7. 111111×11111

8. 101101×10111

9. 1010101×101010

10. 11100101×1001101

11. $110 \div 10$

12. $1000 \div 100$

13. $11000 \div 1000$

14. $1111 \div 11$

15. $111111 \div 1001$

16. $1111001 \div 1011$

17. $10000100 \div 10110$

18. $10001100 \div 100011$

19. $100100001 \div 10001$

20. $1111101000 \div 1100100$

BINARY FRACTIONS

Just as the denary (or decimal) fraction:

$$123{\cdot}476 = 1 \times 10^2 + 2 \times 10^1 + 3 \times 10^\circ + 4 \times 10^{-1} + 7 \times 10^{-2} + 6 \times 10^{-3}$$

or

$$1 \times 100 + 2 \times 10 + 3 \times 1 + 4 \times \tfrac{1}{10} + 7 \times \tfrac{1}{100} + 6 \times \tfrac{1}{1000}$$

so the binary fraction:

$$111{\cdot}111 = 1 \times 2^2 + 1 \times 2^1 + 1 \times 2^\circ + 1 \times 2^{-1} + 1 \times 2^{-2} + 1 \times 2^{-3}$$

or

$$1 \times 4 + 1 \times 2 + 1 \times 1 + 1 \times \tfrac{1}{2} + 1 \times \tfrac{1}{4} + 1 \times \tfrac{1}{8}$$

i.e. 111·111 (base 2) $= 7\frac{7}{8}$ or 7·875 (base 10).

To convert from a denary fraction to its equivalent binary fraction is not quite so simple as the conversion of integers. In simple cases, such as the following, the conversion may be done by inspection:

$$
\begin{aligned}
2{\cdot}5 &= 2\tfrac{1}{2} \text{ (base 10)} && \text{or } 10{\cdot}1 \text{ (base 2)}\\
3{\cdot}75 &= 3+\tfrac{1}{2}+\tfrac{1}{4} && \text{or } 11{\cdot}11 \text{ (base 2)}\\
4{\cdot}875 &= 4+\tfrac{1}{2}+\tfrac{1}{4}+\tfrac{1}{8} && \text{or } 100{\cdot}111 \text{ (base 2)}\\
11{\cdot}0625 &= 11+\tfrac{1}{16} && \text{or (}1011{\cdot}0001 \text{ (base 2)}
\end{aligned}
$$

However, a number such as ·7 cannot be expressed as a finite number of powers of $\frac{1}{2}$. In such a case the best method is to evaluate $7 \div 10$ by long division in the binary scale. The calculation is shown below:

```
            ·1011001100 etc.
     ______________________
1010 | 111·000000000
     | 101 0
     | ______
     |  10 000
     |   1 010
       ______
          1100
          1010
         ______
           10000
            1010
           ______
             1100
             1010
            ______
              1000
```

In this case the non-recurring decimal fraction ·7 is equivalent to the recurring binary fraction $\cdot 10\dot{1}10\dot{0}$.

The processes of division and multiplication of binary fractions involve no new difficulty beyond the necessity for care in the position of the decimal point.

Example (i). Multiply 1101·101 by 11·011 and check the result by performing an equivalent calculation in the denary scale.

We evaluate first of all

$$\begin{array}{r} 1101101 \\ \times\ 11011 \\ \hline 1101101 \\ 11011010 \\ 1101101000 \\ 11011010000 \\ \hline 101101111111 \end{array}$$

Observing that there must be six "bicimal" places in the product of the original number we have:

$$1101{\cdot}101 \times 11{\cdot}011 = 101101{\cdot}111111$$

The equivalent calculation in the denary scale is as follows:

$$13\tfrac{5}{8} \times 3\tfrac{3}{8} = \tfrac{109}{8} \times \tfrac{27}{8} = \tfrac{2943}{64} = 45\tfrac{63}{64}$$

And, of course,

$$\begin{aligned} 101101{\cdot}111111 \text{ (base 2)} &\equiv 45 + \tfrac{1}{2} + \tfrac{1}{4} + \tfrac{1}{8} + \tfrac{1}{16} + \tfrac{1}{32} + \tfrac{1}{64} \\ &\equiv 45\tfrac{63}{64} \text{ (base 10)} \end{aligned}$$

Example (ii). Divide 1·1011101 by 1·101 and check the result by performing an equivalent calculation in the denary scale.

$$\begin{aligned} & 1{\cdot}1011101 \div 1{\cdot}101 \\ = \; & 1101{\cdot}1101 \div 1101 \end{aligned}$$

$$\begin{array}{r|l} & \quad 1{\cdot}0001 \\ \hline 1101 & 1101{\cdot}1101 \\ & 1101{\cdot}1101 \end{array}$$

$$= 1{\cdot}0001$$

The equivalent calculation in the denary scale is as follows:

$$1\tfrac{93}{128} \div 1\tfrac{5}{8} = \frac{\overset{17}{\cancel{221}}}{\underset{16}{\cancel{128}}} \times \frac{\cancel{8}}{\cancel{13}} = 1\tfrac{1}{16}$$

And, of course, 1·0001 (base 2) $= 1\tfrac{1}{16}$ (base 10).

Exercise 5(*e*)

1. Express the following binary fractions as (a) proper fractions in the scale of 10, (b) decimal fractions.

(a) 1·1 (b) 11·01 (c) 101·101
(d) 0·0101 (e) 10·1101 (f) 110·1011
(g) 1010·11011 (h) 1·00011 (i) 10111·00111
(j) 111·111 (k) 1·11111 (l) 11010·011011

2. Express the following denary fractions as binary fractions.

(a) $1\frac{1}{2}$ (b) $2\frac{1}{4}$ (c) $3\frac{1}{8}$
(d) $5\frac{3}{8}$ (e) $7\frac{5}{8}$ (f) $11\frac{11}{16}$
(g) $14\frac{31}{32}$ (h) $17\frac{3}{64}$ (i) $127\frac{127}{128}$

3. Express the following decimal fractions as binary fractions.

(a) 1·25 (b) 3·75 (c) 0·5
(d) 6·875 (e) 7·625 (f) 3·0625
(g) 7·5625 (h) 0·15625 (i) 0·265625

4. Express the following decimal fractions as binary fractions, giving the answers, where necessary, to 5 "bicimal" places.

(a) ·1 (b) ·2 (c) ·3
(d) ·5 (e) ·9 (f) ·13
(g) 2·8875 (h) 3·14 (i) 10·01

5. Simplify the following:

(a) 1·1 × ·1 (b) 10·01 × 1·1
(c) 11·01 × ·11 (d) 1·011 × 1·01
(e) 1·11 × ·011 (f) 10·0101 × 1·101
(g) 1·111 × 11·11 (h) 101·101 × 11·0101
(i) 1·1 ÷ 0·1 (j) 101·1 ÷ ·01
(k) 110·111 ÷ 1·01 (l) 10010·011 ÷ 1·11
(m) 1111·0111 ÷ 1·101 (n) 1·1 ÷ ·00011
(o) 1010·1001 ÷ ·1101 (p) 10000·101111 ÷ 1·0101

USE OF THE BINARY SCALE

For normal purposes the denary scale is certainly more convenient than the binary scale. The latter, however, possesses the great advantage of enabling us to express any number in terms of only two symbols, 0 and 1. Now for 1 we may substitute a current in a circuit, a flash of light or a short audible pulse from a buzzer; 0 is then represented by an absence of current, an unilluminated light filament or a short pause or silence. It is thus possible to perform calculations by using a large number of electrical circuits in which digits and zeros are represented by the presence or absence of electrical currents. This is the principle which underlies the electronic digital computers in use today. So great is the speed at which electric currents flow, and so great is the complexity of modern transistorized computers, that a calculation which might occupy a team of mathematicians in many hours of paperwork can be performed electronically in a fraction of a second.

A fuller appreciation of the principles involved may be gained by constructing a simple binary adder of the type described by Cundy and Rollett in their book *Mathematical Models*. In these small, manually operated devices, the numbers to be added are recorded on banks of switches (on for "1" and off for "0"), and the result of the addition is recorded on a row of flash bulbs, an illuminated bulb representing the digit 1.

Experiments have also been carried out in transmitting signals by Morse apparatus, using a code in which the various letters of the alphabet are represented by various binary numbers. The letters most frequently used are, of course, represented by the simplest binary numbers.

THE BINARY SLIDE RULE

A useful project, which combines activity in the classroom and workshop and which does not involve too much time, is the construction of a simple slide rule. Indeed, the whole job can be carried out quite satisfactorily by most pupils at home. The procedure is as follows:

First construct a table of values of the function 2^x for the values of x: 0, 1, 2, 3, 4. If we are also able to evaluate 2^x for the fractional values $x = \frac{1}{2}, 1\frac{1}{2}, 2\frac{1}{2}, 3\frac{1}{2}$, a much more accurate result will be obtained. These values are as shown below.

x	0	$\frac{1}{2}$	1	$1\frac{1}{2}$	2	$2\frac{1}{2}$	3	$3\frac{1}{2}$	4
2^x	1	1·4	2	2·8	4	5·7	8	11·3	16

Plotting them, a graph similar to that shown in Fig. 1 should be obtained.

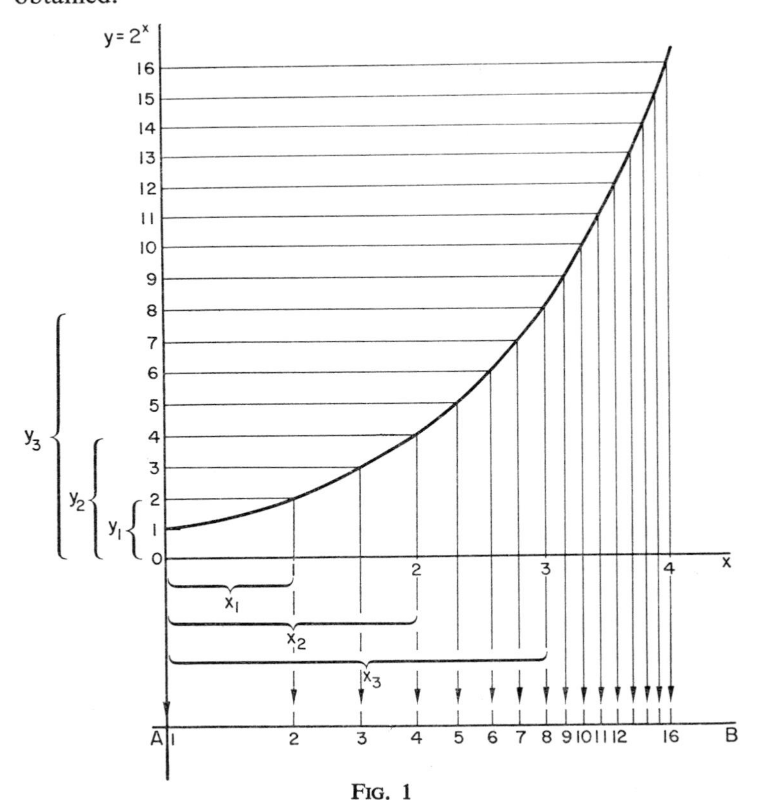

FIG. 1

Now $$2 \times 4 = 8$$

or $$2^1 \times 2^2 = 2^3$$

or $$y_1 \times y_2 = y_3$$

But in performing this multiplication we actually *add the indices*, thus:

$$1+2 = 3$$

or $$x_1+x_2 = x_3$$

Hence, to perform multiplications between numbers on the Y scale, we need only *add* the corresponding numbers on the X scale. It follows that if we re-number the X scale, not with the x values but with the corresponding y values, then the *sum* of the lengths representing any two numbers will be a length representing the *product* of the two numbers. This is the principle underlying the slide rule. We do this as shown on the line AB (Fig. 1).

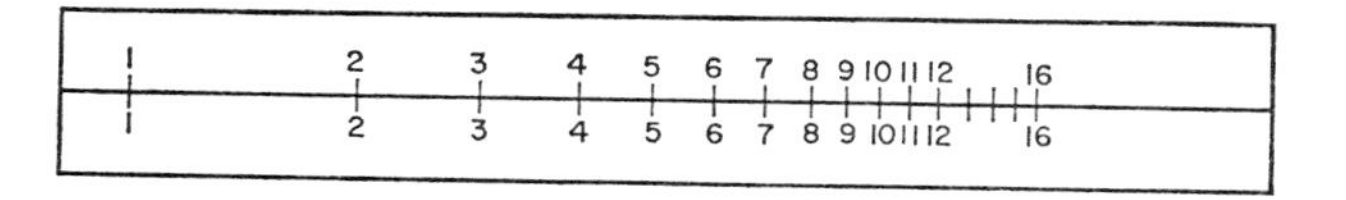

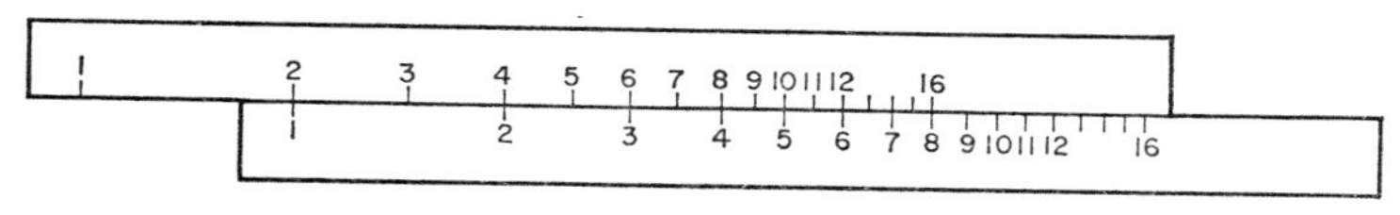

FIG. 2

The scale AB is now transferred in duplicate to two strips of drawing paper as shown in Fig. 2, and when these are made to slide upon each other, the products and quotients of simple numbers may be read off directly. Finally, the scales may be fixed to suitable pieces of wood, dovetail-grooved to slide freely upon each other.

OTHER SCALES OF NOTATION

In our every day calculations we do not always use a scale of ten. In problems involving shillings and pence we do not carry from the pence column until the total is twelve. Addition of minutes and seconds involves a scale of sixty. We even have two common examples of the use of the scale of two—in the addition of pints and quarts, and of stones and quarters.

In this section we shall consider the problem of converting numbers expressed in one base to numbers in a different base. Just as the successive place values in the scale of 10 are:

$$10^3 \quad 10^2 \quad 10^1 \quad 1 \cdot \frac{1}{10} \quad \frac{1}{10^2} \quad \frac{1}{10^3} \quad \ldots$$

and in the scale of 2:

$$2^3 \quad 2^2 \quad 2^1 \quad 1 \cdot \frac{1}{2} \quad \frac{1}{2^2} \quad \frac{1}{2^3} \quad \ldots$$

so our place values in the scale of 3 would be:

$$3^3 \quad 3^2 \quad 3^1 \quad 1 \cdot \frac{1}{3} \quad \frac{1}{3^2} \quad \frac{1}{3^3} \quad \ldots$$

and in the scale of n:

$$n^3 \quad n^2 \quad n^1 \quad 1 \cdot \frac{1}{n} \quad \frac{1}{n^2} \quad \frac{1}{n^3} \quad \ldots$$

Apart from this, the operations of addition, subtraction, multiplication and division are carried out in exactly the same way no matter what base is used.

Example (i). Express the following denary numbers in the scale stated.

(a) 174 in the scale of 6.
(b) 248 in the scale of 5.
(c) 100 in the scale of 3.
(d) $24\frac{19}{49}$ in the scale of 7.
(e) 1475 in the scale of 12.

(a) $174 = 4\times 6^2+5\times 6+0\times 1 = 450$ (base 6)
(b) $248 = 1\times 5^3+4\times 5^2+4\times 5+3\times 1 = 1443$ (base 5)
(c) $100 = 1\times 3^4+2\times 3^2+1\times 1 = 10201$ (base 3)
(d) $24\frac{19}{49} = 3\times 7+3\times 1+2\times\frac{1}{7}+5\times\frac{1}{49} = 33{\cdot}25$ (base 7)
(e) $1475 = \text{ten}\times 12^2+2\times 12+\text{eleven}\times 1$ or t2e (base 12), where t, e are new digits representing "ten" and "eleven" in the denary scale.

Example (ii). The Wroggly people have a place-value system of numbers in which the symbols corresponding to our own numbers are shown below.

×	⊼	⧖	⌧	⊠	⊼×	⊼⊼	⊼⧖	⊼⌧	⊼⊠	...
0	1	2	3	4	5	6	7	8	9	...

Express in the denary scale:

(a) ⌧ ⊼ ⧖ (b) ⊠ × ⌧

Express in Wroggly numbers the denary numbers:

(c) 32 (d) 186
(e) Find the sum of ⌧ ⊼ ⧖ and ⊠ ⌧ ×.

We observe first of all that this number system is essentially a scale of five.

(a) ⌧ ⊼ ⧖ is equivalent to 312 (base 5) or 82 (base 10).
(b) ⊠ × ⌧ is equivalent to 403 (base 5) or 103 (base 10).
(c) 32 (base 10) = 112 (base 5) or ⊼ ⊼ ⧖.
(d) 186 (base 10) = 1221 (base 5) or ⊼ ⧖ ⧖ ⊼.
(e) ⌧ ⊼ ⧖ + ⊠ ⌧ × is equivalent to 312+430 (base 5) or 1242 (base 5) or ⊼ ⧖ ⊠ ⧖ in Wroggly numbers.

Exercise 5(*f*)

1. Express in the scale of 3 the following denary numbers:

(a) 9 (b) 12 (c) 17
(d) 51 (e) 111

2. Express in the scale of 5 the following denary numbers:

(a) 30 (b) 124 (c) 689
(d) 700 (e) 1,000

3. Express in the scale of 6 the following denary numbers:

(a) 18 (b) 40 (c) 96
(d) 250 (e) 1302

4. Express in the scale of 8 the following denary numbers:

(a) 56 (b) 63 (c) 77
(d) 100 (e) 577

5. Express the following denary numbers in the scale indicated:

(a) 17 (base 3) (b) 128 (base 12)
(c) 288 (base 7) (d) 31 (base 4)
(e) 500 (base 9) (f) 362 (base 8)
(g) 42 (base 5) (h) 375 (base 6)
(i) 401 (base 20)

6. Express the following binary numbers in the scale indicated:

(a) 11011 (base 3) (b) 10101 (base 5)
(c) 11101011011 (base 8)

7. Show that by placing weights on either scale pan it is possible to weigh any whole number of pounds using the set of weights 1 lb, 3 lb, 9 lb, 27 lb, 81 lb, etc.

8. A system of numbers

| + 工 王 又 ∋ +| ++ +工 +王 ...

is known to correspond to

0 1 2 3 4 5 6 7 8 9 ...

Express the denary numbers 11 and 94 in this system and convert to the denary scale the numbers ∋+王 and 又王工

Subtract 又|工 from ∋王+

9. After the Great Upheaval, the survivors eventually rediscovered five of our numerals. They used them as follows. Instead of our

	0	1	2	3	4	5	6	7	8	9	...
they wrote	7	4	2	8	3	47	44	42	48	43	...

An examination paper taken by a child in this era contained the questions:

(4) $\begin{array}{r} 873+ \\ 423 \\ \hline \end{array}$ (2) $\begin{array}{r} 32- \\ 84 \\ \hline \end{array}$ (8) $\begin{array}{r} 48 \\ \times 3 \\ \hline \end{array}$ (3) $43 \,\overline{)\, 424}$

Assuming that the signs $+$, $-$, $\times$, $\overline{)}$ have their usual meanings, find what the correct answers should be.

6
GROUPS

You may be rather puzzled by the title of this chapter. Although we shall not start by defining exactly what a group is, a few introductory remarks may help to remove some of the mystery.

Basically, a group is a rather special sort of set. In previous chapters we have quoted many examples of sets in which the elements, although objects of the same kind, have no closer relationship with each other than that. The most that can be said of the set {Mary, Sally, Jane} is that they are all girls' names. We do come across sets, however, which turn out to have rather special properties when we are allowed to add, or multiply, or combine the elements in some specified manner. Later on we shall say what these properties are, and we shall describe sets which possess them as *groups*.

The idea of a group is an extremely important one in many branches of mathematics. By the use of group theory, scientists have been able to make vital progress and discoveries in the fields of crystallography, genetics and quantum mechanics. In mathematics the theory of groups is of great value in the theory of equations, and it was used by the great mathematician Abel to prove that algebraic polynomical equations of degree five or more are in general insoluble.

Much group theory is concerned with the symmetries and patterns which exist in mathematical situations. In this chapter we shall lead up to the definition of a group from some simple aspects of geometrical symmetry.

SYMMETRY AND THE ALPHABET

Let us look at the capital letters of the alphabet. Some of them, such as F and G, are not symmetrical at all. Some, like A and M, are symmetrical, and so are B and K. A and M, however, are symmetrical about a vertical axis, while B and K are symmetrical about a horizontal axis.

Fig. 1

After a rotation through 180°, or a *half-turn about a horizontal axis*, the letters B and K are unaltered. Let us denote this type of rotation or "operation" by the letter p. A and M are unaltered in position and appearance by a *half-turn about the vertical axis*. Let us denote this type of rotation by the letter q. Alternatively, we may think of these operations as *reflections*. The letter A is reflected into itself (or "mapped" into itself) by a thin, double-sided mirror placed along the dotted line in Fig. 1. B is reflected into itself by a mirror placed along the horizontal axis of symmetry.

There is a third type of symmetry associated with the letter N. The operation p on N produces И; the operation q on N produces the same thing. If we follow an operation p on N by an operation q, we do reproduce N. But there is a single operation which will map N into itself; this is a rotation through 180°, or a *half-turn in the plane of the paper* (i.e. about an axis through the midpoint of the oblique stroke, perpendicular to the plane of the paper). Let us denote this type of rotation or operation by r. We shall now write:

$$p(\text{N}) \rightarrow \text{И}$$

So that $$q\,[p(\text{N})]$$

or $$q . p(\text{N}) \rightarrow \text{N}$$

But $$r(\text{N}) \rightarrow \text{N}$$

Hence $$q . p = r$$

In this particular case, $$p . q = r \text{ also,}$$

and indeed, $$p . p \text{ or } p^2 = r.$$

Note that although we write "$q . p$" as though q and p were multiplied together, and although we use an equal sign, the expression $q . p = r$ in this context is to be read as "*operation p followed by operation q produces the same effect as operation r*".

If we look at the alphabet again we find that there are even more symmetrical letters. For example the letter H is unchanged by any of the operations p, q, r. On the other hand, an unsymmetrical letter such as J is changed by any one of the three operations. In order to bring these unsymmetrical letters into our scheme we now introduce a rather trivial operation I, called the *identity* operation. This stands for "leave the letter as it is". Thus, all letters are unchanged by one or other of the operations p, q, r, I. In the case of N, $p . q$ has the same effect as I. Note also that the result of the operation p can be "undone" by a further application of the operation p; $p . p$ or $p^2 = I$. When the effect of an operation x can be reversed by a further operation y, we say that the operation y is the *inverse* of operation x. As far as N is concerned p is its own inverse, and since $q^2 = I$, q is its own inverse; but since $p . q = I$ also, q is another inverse of p. This is to say that in respect of their effects on the letter N, the set of operations I, p, q, r, does not contain *a unique identity element* (r has the same effect as I), and each element does not possess a *unique inverse element* (p and q are both inverses of p itself). We shall see, however, that there are sets of operations applied to certain figures in which uniqueness does exist, and it is such cases which are really interesting.

Exercise 6(*a*)

1. State which capital letters of the alphabet are unaltered by the operation p.

2. State which capital letters of the alphabet are unaltered by the operation q.

3. State which capital letters of the alphabet are unaltered by the operation r.

4. Which capital letters of the alphabet are unaltered by all three of the operations p, q, r?

5. Draw the letter J as it appears after the application of each of the following operations:

(a) p	(b) q	(c) r
(d) I	(e) p^2	(f) q^2
(g) r^2	(h) $p.q$	(i) $q.p$
(j) $r.p$	(k) $p.r$	(l) $q.r$
(m) $r.q$	(n) $I.p$	(o) $I.q$
(p) $I.r$	(q) I^2	

State which of the single operations p, q, r or I is the equivalent of (e)–(q) above. Now complete the following "multiplication" table:

	×	I	p	q	r	first operation
	I	.	.	.	.	
second	p	.	.	.	.	
operation	q	.	.	.	.	
	r	.	.	.	.	

GROUPS IN GEOMETRY. THE GROUP OF THE RECTANGLE

What we are really going to investigate now is the *group of movements*, or the *group of symmetries*, of the rectangle. As we shall see, this *group of movements* also forms a *group* in the mathematical sense; hence the term "the group of the rectangle".

Consider now the problem of fitting a rectangular pane of glass into a rectangular window frame of the same dimensions. In how many ways can we place the glass in position?

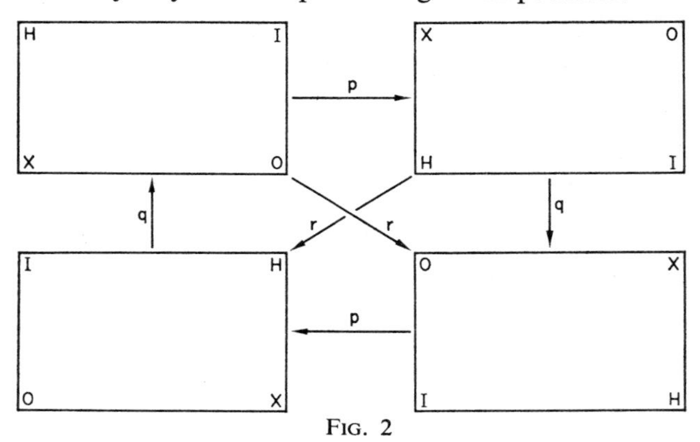

FIG. 2

If we take a rectangle of glass, or celluloid, or clear acetate paper, and label it H, I, O, X, we find that there are four possible ways of fitting it into a given frame of the same size. These are shown in Fig. 2, the arrows indicating the operations by which one position may be obtained from another.

You should check that these linking operations are in fact correct. It is also worth noting the highly suitable system of lettering; H, I, O and X are all unaltered by any one of the operations I, p, q and r. Further, the connections between the linking operations are as follows:

$$p^2 = q^2 = r^2 = I; \quad p.q = q.p = r;$$
$$p.r = r.p = q; \quad q.r = r.q = p; \quad \text{and, of course,}$$
$$I^2 = I, \quad I.p = p, \quad I.q = q \quad \text{and} \quad I.r = r$$

We can see the situation at a glance if we set these results out in the form of a multiplication table:

	×	I	p	q	r	first operation
	I	I	p	q	r	
second	p	p	I	r	q	
operation	q	q	r	I	p	
	r	r	q	p	I	

Notice that under our rule of "multiplication", where $p.q$ means "operation q followed by operation p", the set of operations $\{I, p, q, r\}$ has the following properties:

1. The "product" of any two elements of the set is also an element of the set. This is sometimes called the "*closure property*". Alternatively we may say that the rule of "multiplication" is a *binary operation.*

2. There is only one element, namely I, which, when multiplying or multiplied by any element of the set, leaves that element unaltered. i.e.

$$I.I = I, \quad I.p = p.I = p, \quad I.q = q.I = q, \quad I.r = r.I = r$$

We say that the set has a *unique identity element I.*

3. Each element of the set has *a unique inverse element.* In fact in this case each element is its own inverse. The effect of the operation p is cancelled by a further operation p, i.e. $p.p$ or $p^2 = I$. Likewise $I^2 = I$, $q^2 = I$ and $r^2 = I$.

4. *The associative law holds.* That is to say that any expression of the type $a \times (b \times c)$ is equal to $(a \times b) \times c$.

e.g. $(q \times r) = p \quad \therefore \ p \times (q \times r) = p \times p = I$

Also $(p \times q) = r \quad \therefore \ (p \times q) \times r = r \times r = I$

and so on for all possible trios of elements.

Any set in which there is a rule for multiplying or combining the elements such that the conditions 1, 2, 3 and 4 apply, is called a *group*.

Actually, the group of the rectangle has an additional property which we have not mentioned, and which indeed need not necessarily be possessed by a group. This is the *commutative property*. An algebra in which every two elements a, b, obey the rule $a \times b = b \times a$ is said to be *commutative*. Ordinary arithmetic is commutative under multiplication and addition; for example:

$$2 \times 3 = 3 \times 2 = 6$$

and $$2+3 = 3+2 = 5$$

Is it commutative under division and subtraction?

The algebra of sets is also commutative:

$$A \cap B = B \cap A$$

and $$A \cup B = B \cup A$$

In the case of I, p, q, r we saw that $p.q = q.p$, $p.r = r.p$, etc. We shall see later, however, that many algebras are not commutative, and generally groups do not exhibit this commutative property. Special groups, like the group of the rectangle, which are commutative, are called *commutative groups* or *Abelian groups*.

[*Note.* In Question 5 of Ex. 6(a) you should have obtained the group of the rectangle—or of a rectangular sheet bearing the letter J. This was not the group of the letter J, which itself has no axes of symmetry.]

Exercise 6(*b*)

1.

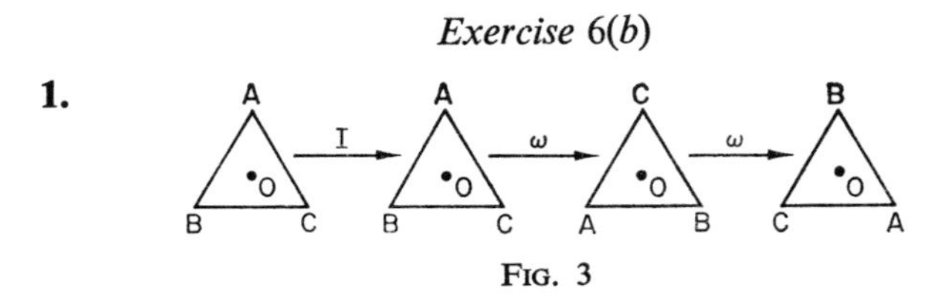

FIG. 3

ABC is an equilateral triangle and Fig. 3 shows some of the possible ways in which it may be placed in a triangular frame of the same size. The operation ω means "rotate the triangle anti-clockwise in its own plane about the centre of gravity O through an angle of 120°". Complete the multiplication table:

$\times$	1	ω	ω^2
1	.	.	.
ω	.	.	.
ω^2	.	.	.

and show that the operations $\{I, \omega, \omega^2\}$ form a commutative group.

[This group is called the "group of rotations of the equilateral triangle".]

2. Fig. 3 does not show all the possible ways of placing the $\triangle$ ABC in position; there are three other arrangements:

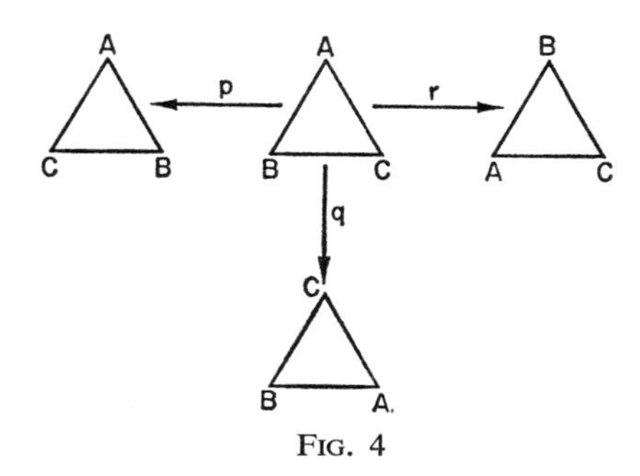

FIG. 4

These result from the following operations:

A half-turn about the perpendicular from A to BC .
A half-turn about the perpendicular from B to AC .
A half-turn about the perpendicular from C to AB .

Thus, for example,

p followed by q, or ω^2 gives B
i.e. $q.p = \omega^2$ C A

q followed by p, or ω gives C
i.e. $p.q = \omega$ A B

Investigate all other "products" of the elements of the set of operations $\{I, \omega, \omega^2, p, q, r\}$, completing the "multiplication" table shown below as you do so.

	$\times$	I	ω	ω^2	p	q	r	first operation
	I	.	.	.	.	.	.	
	ω	.	ω^2	.	.	.	.	
second	ω^2	.	.	.	.	.	.	
operation	p	.	.	.	.	ω	.	
	q	.	.	.	ω^2	.	.	
	r	.	.	.	.	.	.	

Show that the product of any two elements is also an element of the set; that the rule of combination is associative; that the element I is unique, and that each element has one and only one inverse. In other words show that the operations $\{I, \omega, \omega^2, p, q, r\}$ form a non-commutative group. This group is the full group of the equilateral triangle.

3.

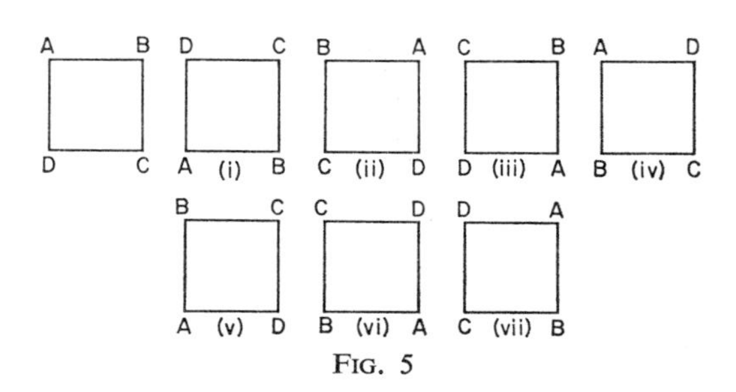

FIG. 5

A square pane of glass ABCD can be placed into a square frame of the same size in the eight different ways shown in Fig. 5. The seven changes (i)–(vii) from the original position result from the following operations:

(i) A half-turn about a horizontal axis . . . . operation p

(ii) A half-turn about a vertical axis. . . . . . operation q

(iii) A half-turn about a diagonal DB (↻). . . operation r

[*Note.* This definition of r is different from the previous one.]

(iv) A half-turn about a diagonal AC (⤡) . . . operation s

(v) An anti-clockwise rotation through one right angle . . . . operation ω

(vi) An anti-clockwise rotation through two right angles . . . . operation ω^2

(vii) An anti-clockwise rotation through three right angles . . . . operation ω^3

The original figure may be regarded as having turned through an anti-clockwise rotation of zero angle—the identity operation—1.

Note. This definition of ω is also different from the previous one. Further, the identity operation 1 is clearly equivalent to an anti-clockwise rotation of four right angles, i.e. $\omega^4 = 1$.

(a) Consider first the operations $\{1, \omega, \omega^2, \omega^3\}$. As in the previous examples, construct a "multiplication" table and show that these operations form a commutative group.

(b) Now check the entries given and complete the "multiplication" table of the operations $\{p, q, r, s\}$.

	×	p	q	r	s	first operation
	p	1	ω^2	ω^3	ω	
second operation	q	.	1	ω	ω^3	
	r	.	.	1	ω^2	
	s	.	.	.	1	

Does $\{p, q, r, s\}$ form a group?

(c) Check the first multiplication table given and complete the second.

	×	1	ω	ω^2	ω^3	first operation
	p	p	s	q	r	
second operation	q	q	r	p	s	
	r	r	p	s	q	
	s	s	q	r	p	

	×	p	q	r	s	first operation
	1	.	.	.	.	
second	ω	.	.	.	.	
operation	ω^2	.	.	.	.	
	ω^3	.	.	.	.	

(d) Finally, using your results in (a), (b) and (c), complete the multiplication table for the whole set of operations $\{1, \omega, \omega^2, \omega^3, p, q, r, s\}$.

	×	1	ω	ω^2	ω^3	p	q	r	s	first operation
	1	.	.	.	.	.	.	.	.	
second	ω	.	.	.	.	.	.	.	.	
operation	ω^2	.	.	.	.	.	.	.	.	
	ω^3	.	.	.	.	.	.	.	.	
	p	p	s	q	r	1	ω^2	ω^3	ω	
	q	q	r	p	s	.	1	ω	ω^3	
	r	r	p	s	q	.	.	1	ω^2	
	s	s	q	r	p	.	.	.	1	

If you have done this correctly, you should be able to show that the set of operations $\{1, \omega, \omega^2, \omega^3, p, q, r, s\}$ is a group. It is called the group of the square.

GROUPS IN ARITHMETIC

Counting in the scale of 10 we have:

1, 2, 3, 4, 5, 6, 7, 8, 9, 10, 11, 12 . . .

Counting in the scale of 5, however, we proceed:

1, 2, 3, 4, 10, 11, 12, 13, 14, 20, 21, 22 . . .

If we now retain only the units column we have what are called the *integers modulo* 5:

1, 2, 3, 4, 0, 1, 2, 3, 4, 0, 1, 2 . . .

A simpler way of putting this is to say that we have divided the natural numbers in turn by 5 and have kept only the remainders.

In this curious "modulo" arithmetic we have such results as:

$$2+3 = 0$$
$$2+4 = 1$$
$$3+4 = 2$$
$$4+4 = 3$$
etc.

The complete addition table is shown below:

+	0	1	2	3	4
0	0	1	2	3	4
1	1	2	3	4	0
2	2	3	4	0	1
3	3	4	0	1	2
4	4	0	1	2	3

Notice that in the set $\{0, 1, 2, 3, 4\}$ mod. 5, and using ordinary arithmetical addition:

(i) The sum of any two elements is also an element of the set.

(ii) There is an identity element, i.e. an element which does not alter the element upon which it operates. This element is 0, for $0+1 = 1+0 = 1$; $0+2 = 2+0 = 2$, etc.

(iii) For each element there is one, and only one, element (its inverse) which cancels its effect. Thus, starting with zero and adding any number, say 3, we can find another number, 2, which on addition brings us back to zero, for $3+2 = 0$. Furthermore, this is the only number which will do this. Likewise, the inverse of 4 is 1, for $4+1 = 0$, and the inverse of 0 is, of course, 0.

(iv) Finally, the rule of combining three elements, here normal addition, is associative. For any three numbers we can always write $(a+b)+c = a+(b+c)$.

Now (i)–(iv) are the properties of a group. Hence the integers modulo 5, under addition, form a group. Actually the whole set of positive and negative integers form a group as well. It is impossible to list the set unless we write $\{\ldots -3, -2, -1, 0, 1, 2, 3, 4 \ldots\}$, and it is also impossible to write out the addition table. It is an *infinite group*. Nevertheless:

(i) the sum of any two integers is another integer,
(ii) the identity element is 0,
(iii) each element, say a, has only one inverse, $(-a)$,
(iv) addition is associative.

Let us return to our integers modulo 5 and see what happens if we multiply them together in the common arithmetical sense.

×	0	1	2	3	4
0	0	0	0	0	0
1	0	1	2	3	4
2	0	2	4	1	3
3	0	3	1	4	2
4	0	4	3	2	1

One glance at this tells us that we do *not* have a group, for the identity element must be 1, as it is the only element which leaves any element unaltered under multiplication; $0 \times 1 = 0$, $2 \times 1 = 2$, $4 \times 1 = 4$, etc. But in this case the element 0 has no inverse, therefore the properties of a group are not satisfied. If, however, we exclude the element 0, we find that the integers $\{1, 2, 3, 4\}$ modulo 5 *do* form a group under multiplication.

Exercise 6(*c*)

Test and state which of the following sets are groups under addition.

1. The integers modulo 3.
2. The integers modulo 6.
3. The integers modulo 8.
4. The positive rational (fractional) numbers.
5. The negative integers.
6. The real numbers.
7. The subsets of a given universal set under union.
8. The subsets of a given universal set under intersection.

Test and state which of the following sets are groups under multiplication.

9. The integers modulo 3 (excluding 0).

10. The integers modulo 4 (excluding 0).

11. The integers modulo 6 (excluding 0).

12. The integers modulo 7 (excluding 0).

13. The integers modulo 8 (excluding 0).

(Can you see any general rule that might apply as a result of your answers to Questions 9–13?)

14. The integral powers of 2 $\left\{\ldots 2^3, 2^2, 2^1, 2^0, \frac{1}{2}, \frac{1}{2^2}, \frac{1}{2^3} \ldots\right\}$.

15. The positive rational numbers.

16. The real numbers.

17. The real numbers excluding 0.

GROUPS IN ALGEBRA

Certain sets of functions form groups under a rule by which we *substitute* one function in another. Most of these are beyond the scope of this book, but there is one simple example which we shall pursue in the next section.

Exercise 6(*d*)

Consider the four functions $f_1(x) = x$, $f_2(x) = -x$, $f_3(x) = \frac{1}{x}$,

$$f_4(x) = -\frac{1}{x}.$$

By "$f_2 f_3$" we mean $f_2[f_3(x)]$.

Now f_3 "acts" on x to produce $\frac{1}{x}$,

or on 2 to produce $\frac{1}{2}$.

Alternatively f_3 maps x into $\frac{1}{x}$

and maps 2 into $\frac{1}{2}$.

Also, f_2 maps x into $-x$,

or it maps $\frac{1}{x}$ into $-\frac{1}{x}$.

So $$f_2[f_3(x)] = f_2\left(\frac{1}{x}\right) = -\frac{1}{x} = f_4(x)$$

or $$f_2 f_3 = f_4.$$

Now let us look at $f_4 f_2$.

$$f_4 f_2 = f_4[f_2(x)] = f_4(-x) = -\frac{1}{-x} = \frac{1}{x} = f_3(x)$$

$$\therefore f_4 f_2 = f_3$$

Express each of the following as a single function.

1. $f_2 f_4$	**2.** $f_3 f_2$	**3.** $f_3 f_3$
4. $f_3 f_4$	**5.** $f_4 f_3$	**6.** $f_4 f_4$
7. $f_1 f_1$	**8.** $f_1 f_2$	**9.** $f_1 f_3$
10. $f_1 f_4$	**11.** $f_4 f_1$	**12.** $f_2 f_2$

Now complete the "multiplication" table below and show that the set of functions $\{f_1, f_2, f_3, f_4\}$ form a commutative group. Which is the identity element?

	f_1	f_2	f_3	f_4
f_1	.	.	.	.
f_2	.	.	.	.
f_3	.	.	.	.
f_4	.	.	.	.

SUBGROUPS

We have seen that a set containing more than one element has proper subsets, and also that a group is a set. Is it true to say that if a set is a group then its subsets form groups also? In other words, is a subset of a group always a *subgroup*? It is not difficult to see that this is not so. Consider once again our integers modulo {0, 1, 2, 3, 4}. One subset is {1, 2, 3}. Under addition $2+3 = 0$, and under multiplication $2 \times 2 = 4$, and neither 0 nor 4 are members of the set. Hence the subset {1, 2, 3} is neither a group under addition nor under multiplication. In fact this group has no subgroups at all.

Now let us look at the group of the rectangle:

$\times$	I	p	q	r
I	I	p	q	r
p	p	I	r	q
q	q	r	I	p
r	r	q	p	I

We can extract from this table the following subsidiary tables:

$\times$	I
I	I

The trivial group $\{I\}$. (The group of any letter having no symmetry, e.g. the letter J.)

$\times$	I	p
I	I	p
p	p	I

The group $\{I, p\}$. (The group, for example, of the letter B which has symmetry about one axis—the horizontal axis.)

$\times$	I	q
I	I	q
q	q	I

The group $\{I, q\}$. (The group, for example, of the letter A, which has symmetry about one axis—the vertical axis.)

$\times$	I	r
I	I	r
r	r	I

The group $\{I, r\}$. (The group, for example, of the letters N and S, which have "half-turn symmetry".)

The rectangle possesses all these symmetries so, as we might expect, its group possesses all these individual symmetries in the form of subgroups.

Exercise 6(*e*)

1. Examine the multiplication table of the group $\{f_1, f_2, f_3, f_4\}$ obtained in Ex. 6(d) and list all its subgroups.

2. Show that the group of the square $\{1, \omega, \omega^2, \omega^3, p, q, r, s\}$ obtained in Ex. 6(b), Question 3 contains the subgroup of the rectangle. List all its other subgroups.

3. Show that the group of the equilateral triangle $\{I, \omega, \omega^2, p, q, r\}$ (Ex. 6(b), Question 2) contains the subgroup of rotations in the plane $\{I, \omega, \omega^2\}$ (Ex. 6(b), Question 1). Does it contain any other non-trivial subgroups?

4. State two subgroups of the group of real numbers under addition.

5. State two subgroups of the group of real numbers (excluding 0) under multiplication.

6. The set of integers (excluding 0) is not a group under multiplication. Find a subset of it which is a group.

7. Form the addition table of the group of integers $\{0, 1, 2, 3, 4, 5\}$ modulo 6 and show that it has subgroups of order 2 and order 3. List these. [Notice that 2 and 3 are factors of 6.]

8. Show that the integers $\{1, 3, 5, 7\}$ modulo 8 form a group under multiplication. List its three subgroups of order 2.

ISOMORPHIC GROUPS

We sometimes find, after abstracting the characteristic groups of symmetries possessed by two quite different sets of elements, that the two groups reveal a remarkable similarity of structure.

The discovery of links such as this can pave the way for further advances, and this is one of the valuable aspects of group theory.

As an example, consider the group of rotations of the equilateral triangle $\{1, \omega, \omega^2\}$ and the integers modulo 3 under addition. Their multiplication and addition tables respectively are:

$\times$	1	ω	ω^2
1	1	ω	ω^2
ω	ω	ω^2	1
ω^2	ω^2	1	ω

$\times$	0	1	2
0	0	1	2
1	1	2	0
2	2	0	1

If, in the first group, we replace 1 by 0, ω by 1 and ω^2 by 2, we obtain the second group. In other words, the two groups have *essentially* the same structure. To use a Greek word which means the same thing, we say that these groups are *isomorphic* under the mapping $1 \leftrightarrow 0$, $\omega \leftrightarrow 1$, $\omega^2 \leftrightarrow 2$.

Exercise 6(*f*)

1. Show that the group of operations $\{1, \omega, \omega^2, \omega^3\}$ obtained in Ex. 6(b), Question 3(a) and the group of integers $\{1, 2, 4, 3\}$ modulo 5 under multiplication are isomorphic.

2. Show that the group of functions $\{f_1, f_2, f_3, f_4\}$ obtained in Ex. 6(d) is isomorphic with the group of prime integers $\{1, 3, 5, 7\}$ modulo 8 under multiplication.

3. Show that the group of successive rotations $\{1, \omega, \omega^2, \omega^3, \omega^4, \omega^5\}$, where $\omega^6 = 1$, is isomorphic with the group of integers $\{0, 1, 2, 3, 4, 5\}$ modulo 6 under addition.

4. Can you generalize the result of Question 3?

5. Show that the group of integers modulo 8 under addition has a subgroup of order 4, and that this in turn has a subgroup of order 2.

6. With reference to Question 2, list the subgroups of the prime integers {1, 3, 5, 7} modulo 8 and list the subgroups of $\{f_1, f_2, f_3, f_4\}$ to which these are respectively isomorphic.

7.

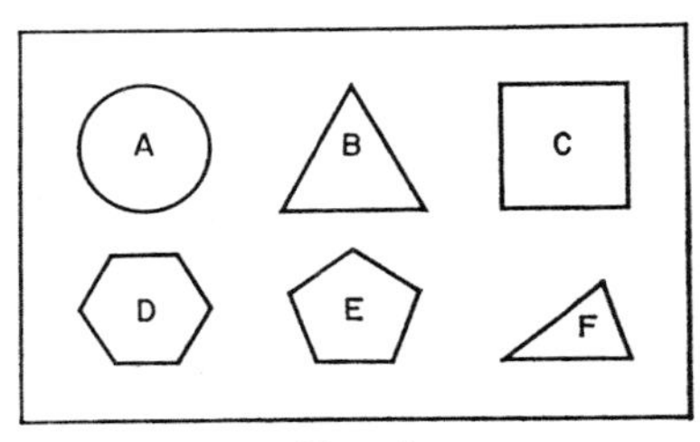

FIG. 6

Fig. 6 shows a nursery toy in which blocks are "posted" through holes of the same shape. Place, in order of difficulty, the operations of "posting" through A, B, C, D, E, F. Explain your answer in terms of group theory. Are the groups of symmetries associated with any of these shapes isomorphic? Is any one of these groups a subgroup of another?

7
MATRICES

A MATRIX IS AN OPERATOR

Let us take another look at the problem, which we discussed in Chapter 6, of placing a rectangular sheet of glass into a frame of the same dimensions. We discovered that this could be done in four different ways. We can place it in position as shown in Fig. 1 (i); secondly, we can rotate it through 180° about its horizontal axis of symmetry (operation p) and place it as shown in (ii); third, we can rotate it through 180° about its vertical axis of symmetry (operation q) and place it as shown in (iii); finally, we can rotate it through 180° in its own plane about an axis through 0 (operation r) and place it as shown in (iv).

Now the operation p is equivalent to "sending" the point A_0, or $(x_0 y_0)$, to the new position A_1, or $(x_1 y_1)$, where

$$\begin{aligned} x_1 &= x_0 \\ y_1 &= -y_0 \end{aligned} \qquad \text{(I)}$$

or, in full,

$$\begin{aligned} x_1 &= 1.x_0+0.y_0 \\ y_1 &= 0.x_0-1.y_0. \end{aligned}$$

Writing this still more compactly in the form

$$\begin{pmatrix} x_1 \\ y_1 \end{pmatrix} = \begin{pmatrix} 1 & 0 \\ 0 & -1 \end{pmatrix} \cdot \begin{pmatrix} x_0 \\ y_0 \end{pmatrix} \qquad \text{(II)}$$

we can think of $\begin{pmatrix} 1 & 0 \\ 0 & -1 \end{pmatrix}$ "operating" on the point $\begin{pmatrix} x_0 \\ y_0 \end{pmatrix}$ to produce the new point $\begin{pmatrix} x_1 \\ y_1 \end{pmatrix}$. We say that $\begin{pmatrix} 1 & 0 \\ 0 & -1 \end{pmatrix}$ *transforms* the point $\begin{pmatrix} x_0 \\ y_0 \end{pmatrix}$, or A_0, into the point $\begin{pmatrix} x_1 \\ y_1 \end{pmatrix}$, or A_1. So what p did in the last chapter, $\begin{pmatrix} 1 & 0 \\ 0 & -1 \end{pmatrix}$ does in the Cartesian plane XOY.

The expression $\begin{pmatrix} 1 & 0 \\ 0 & -1 \end{pmatrix}$ is called a *matrix*. A *matrix* is simply an array of numbers or elements. The following are all matrices:

$$\begin{pmatrix} a & b \\ c & d \\ e & f \end{pmatrix}, \quad \begin{pmatrix} 1 & 2 & 3 \\ 4 & 0 & -2 \end{pmatrix}, \quad \begin{pmatrix} x_0 \\ y_0 \end{pmatrix}, \quad (p,q,r), \quad \begin{pmatrix} 1 & 2 \\ 3 & 4 \end{pmatrix}.$$

(i) (ii) (iii) (iv) (v)

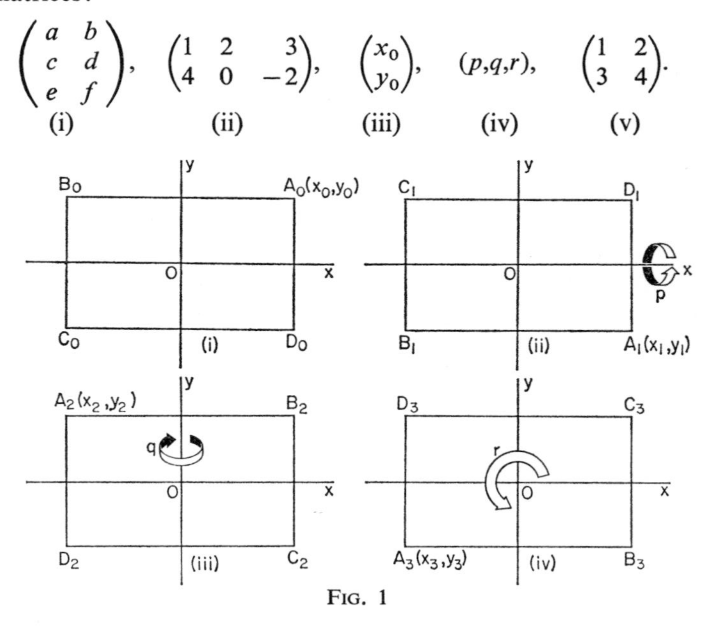

FIG. 1

In describing a matrix we speak of its *order*. An "$l \times m$" matrix is one which has l *rows* and m *columns*.

Thus (i) is a 3×2 matrix,
(ii) is a 2×3 matrix,
(iii) is a 2×1 matrix or *column matrix of order* 2,
(iv) is a 1×3 matrix or *row matrix of order* 3,
(v) is a 2×2 matrix or a *square matrix of order* 2.

A matrix has no numerical value but may act as an *operator*. $\begin{pmatrix} 1 & 0 \\ 0 & -1 \end{pmatrix}$ performs the operation p. We will therefore call it p for the time being.

In this chapter we shall restrict our attention mainly to 2×2 matrices and second order row and column matrices.

Exercise 7(*a*)

Try these questions yourself before reading the rest of this section. In each case refer to Fig. 1 and proceed as follows:

(i) Write down the equations required (as in equation I).
(ii) Write these equations in matrix form (as in equation II).
(iii) Finally state simply the matrix which effects the required transformation.

1. Find the matrix which transforms the point A_0 into the point A_2.

2. Find the matrix which transforms the point A_0 into the point A_3.

3. Find the matrix which transforms the point A_2 into the point A_3.

4. Find the matrix which transforms the point A_1 into the point A_3.

5. Find the matrix which transforms the point A_0 into itself.

Now let us see if you have answered the questions in Ex. 7(a) correctly.

The point A_0 is transformed into the point A_2 by means of the following substitutions:

$$x_2 = -x_0$$
$$y_2 = y_0$$

or

$$x_2 = -x_0 + 0y_0$$
$$y_2 = 0x_0 + y_0$$

or

$$\begin{pmatrix} x_2 \\ y_2 \end{pmatrix} = \begin{pmatrix} -1 & 0 \\ 0 & 1 \end{pmatrix} \cdot \begin{pmatrix} x_0 \\ y_0 \end{pmatrix}$$

Thus, the matrix $\begin{pmatrix} -1 & 0 \\ 0 & 1 \end{pmatrix}$ transforms the point $\begin{pmatrix} x_0 \\ y_0 \end{pmatrix}$, or A_0, into the point $\begin{pmatrix} x_2 \\ y_2 \end{pmatrix}$, or A_2.

Now this is precisely the effect of the operation q, so let us denote $\begin{pmatrix} -1 & 0 \\ 0 & 1 \end{pmatrix}$ by q.

To transform A_0 into A_3 we have:

$$x_3 = -x_0$$
$$y_3 = \quad -y_0$$

or
$$\begin{pmatrix} x_3 \\ y_3 \end{pmatrix} = \begin{pmatrix} -1 & 0 \\ 0 & -1 \end{pmatrix} \begin{pmatrix} x_0 \\ y_0 \end{pmatrix}$$

Now this is the effect of the operation r, so we denote $\begin{pmatrix} -1 & 0 \\ 0 & -1 \end{pmatrix}$ by r.

The matrix which transforms A_2 into A_3 is, of course, the same as the matrix which transforms A_0 into A_1, i.e. p or $\begin{pmatrix} 1 & 0 \\ 0 & -1 \end{pmatrix}$.

The matrix which transforms A_1 into A_3 is the same as that which transforms A_0 to A_2, i.e. q or $\begin{pmatrix} -1 & 0 \\ 0 & 1 \end{pmatrix}$.

To transform A_0 into itself we have

$$x_0 = x_0$$
$$y_0 = \quad y_0$$

or
$$\begin{pmatrix} x_0 \\ y_0 \end{pmatrix} = \begin{pmatrix} 1 & 0 \\ 0 & 1 \end{pmatrix} \begin{pmatrix} x_0 \\ y_0 \end{pmatrix}$$

This matrix, which leaves any point unaltered, has exactly the same effect as the identity operation I. We therefore denote $\begin{pmatrix} 1 & 0 \\ 0 & 1 \end{pmatrix}$ by I.

Since, in matrix algebra, $\begin{pmatrix} 1 & 0 \\ 0 & 1 \end{pmatrix}$ has the same effect as the symbol 1, or unity, in ordinary algebra, we call it the *unit matrix*. There is also another interesting matrix $\begin{pmatrix} 0 & 0 \\ 0 & 0 \end{pmatrix}$. You will see that this sends any point, including A_0, A_1, A_2 and A_3, to the origin (0,0). We therefore call it the *null matrix*. Clearly, it has a similar effect to that of 0 or *zero* in ordinary algebra.

MULTIPLICATION OF MATRICES

In the last chapter we saw that operation p followed by operation q has the same effect as operation r, i.e. $q.p = r$. We should expect, therefore, that the effect of matrix p followed by the effect of matrix q is the same as the effect of matrix r. In other words, we should be able to write:

$$\begin{pmatrix} -1 & 0 \\ 0 & 1 \end{pmatrix} \cdot \begin{pmatrix} 1 & 0 \\ 0 & -1 \end{pmatrix} = \begin{pmatrix} -1 & 0 \\ 0 & -1 \end{pmatrix}$$

Another way of looking at this is to say that we can "send" A_0 to A_3 either (a) via A_1, or (b) directly.

Now $$\begin{pmatrix} x_1 \\ y_1 \end{pmatrix} = \begin{pmatrix} 1 & 0 \\ 0 & -1 \end{pmatrix} \cdot \begin{pmatrix} x_0 \\ y_0 \end{pmatrix}$$

and $$\begin{pmatrix} x_3 \\ y_3 \end{pmatrix} = \begin{pmatrix} -1 & 0 \\ 0 & 1 \end{pmatrix} \cdot \begin{pmatrix} x_1 \\ y_1 \end{pmatrix}$$

$$\therefore \begin{pmatrix} x_3 \\ y_3 \end{pmatrix} = \begin{pmatrix} -1 & 0 \\ 0 & 1 \end{pmatrix} \cdot \begin{pmatrix} 1 & 0 \\ 0 & -1 \end{pmatrix} \cdot \begin{pmatrix} x_0 \\ y_0 \end{pmatrix}$$

But $$\begin{pmatrix} x_3 \\ y_3 \end{pmatrix} = \begin{pmatrix} -1 & 0 \\ 0 & -1 \end{pmatrix} \cdot \begin{pmatrix} x_0 \\ y_0 \end{pmatrix} \text{ directly,}$$

hence $$\begin{pmatrix} -1 & 0 \\ 0 & 1 \end{pmatrix} \cdot \begin{pmatrix} 1 & 0 \\ 0 & -1 \end{pmatrix} = \begin{pmatrix} -1 & 0 \\ 0 & -1 \end{pmatrix} \text{ as before.}$$

The question now is, how is this multiplication carried out?

If we consider a more general case it will help us to see the rule.

Suppose
$$x_2 = Ax_1 + By_1$$
$$y_2 = Cx_1 + Dy_1$$

and
$$x_1 = ax_0 + by_0$$
$$y_1 = cx_0 + dy_0$$

Then
$$x_2 = A(ax_0 + by_0) + B(cx_0 + dy_0)$$
$$y_2 = C(ax_0 + by_0) + D(cx_0 + dy_0)$$

i.e.
$$x_2 = (Aa + Bc)x_0 + (Ab + Bd)y_0$$
$$y_2 = (Ca + Dc)x_0 + (Cb + Dd)y_0$$

i.e.
$$\begin{pmatrix} A & B \\ C & D \end{pmatrix} \cdot \begin{pmatrix} a & b \\ c & d \end{pmatrix} = \begin{pmatrix} Aa+Bc & Ab+Bd \\ Ca+Dc & Cb+Dd \end{pmatrix}$$

or
$$L \, . \, M = N$$

Note. The element in the first row, first column of N, is the sum of the products of corresponding elements in the first row of L and the first column of M.

1st row, 1st col.
element

i.e. $A \quad B$ (1st row) $\quad \begin{matrix} a \\ b \end{matrix}$ } 1st col. $\rightarrow$ $Aa + Bd$

or "Left by top plus right by bottom"

In fact every element in the "product" N is formed in this way.

Thus 2nd row $C \quad D$ $\quad \begin{matrix} a \\ c \end{matrix}$ 1st col. $\rightarrow$ $Ca + Dc$

2nd row, 1st col.
element.

or "left by top plus right by bottom"

Check the other two elements of N for yourself.

We now see that:

$$\begin{pmatrix} -1 & 0 \\ 0 & 1 \end{pmatrix} \cdot \begin{pmatrix} 1 & 0 \\ 0 & -1 \end{pmatrix} = \begin{pmatrix} (-1)(1)+(0)(0) & (-1)(0)+(0)(-1) \\ (0)(1)+(1)(0) & (0)(0)+(1)(-1) \end{pmatrix}$$

$$= \begin{pmatrix} -1 & 0 \\ 0 & -1 \end{pmatrix}$$

If you refer to the multiplication table of the group $\{I, p, q, r\}$ in Chapter 6, you will recall that this was a commutative group, i.e.

$$q.p = p.q = r.$$

Likewise we find that:

$$\begin{pmatrix} -1 & 0 \\ 0 & 1 \end{pmatrix} \cdot \begin{pmatrix} 1 & 0 \\ 0 & -1 \end{pmatrix} = \begin{pmatrix} 1 & 0 \\ 0 & -1 \end{pmatrix} \cdot \begin{pmatrix} -1 & 0 \\ 0 & 1 \end{pmatrix} = \begin{pmatrix} -1 & 0 \\ 0 & -1 \end{pmatrix}$$

This is not true of matrices in general. Looking at L and M again we have:

$$\begin{pmatrix} A & B \\ C & D \end{pmatrix} \cdot \begin{pmatrix} a & b \\ c & d \end{pmatrix} = \begin{pmatrix} Aa+Bc & Ab+Bd \\ Ca+Dc & Cb+Dd \end{pmatrix}$$

whereas

$$\begin{pmatrix} a & b \\ c & d \end{pmatrix} \cdot \begin{pmatrix} A & B \\ C & D \end{pmatrix} = \begin{pmatrix} Aa+Cb & Ba+Db \\ Ac+Cd & Bc+Dd \end{pmatrix}$$

and the two products are quite different.

[*Note*. We can use the above rule to multiply together any two matrices A and B. The element in the rth row and s^{th} column of the product is the sum of the products of corresponding elements in the rth row of A and the sth column of B. For this to be possible, of course, there must be as many columns in the matrix A as there are rows in the matrix B. If A is an $l \times m$ matrix and B is an $m \times n$ matrix, AB is an $l \times n$ matrix. However, $B.A$ cannot be worked out unless $l = n$, and in this case $B.A$ would be a square matrix of order m. Thus, not only may $A.B$ and $B.A$ have different elements; they may actually be matrices of quite different orders.]

Exercise 7(*b*)

1. If $A = \begin{pmatrix} 1 & 2 \\ 0 & 3 \end{pmatrix}$, $B = \begin{pmatrix} 2 & 1 \\ 4 & 3 \end{pmatrix}$, form the matrices $A.B$ and $B.A$.

2. If $C = \begin{pmatrix} 2 & -1 \\ 1 & 2 \end{pmatrix}$, $D = \begin{pmatrix} 3 & 2 \\ 2 & 1 \end{pmatrix}$, form the matrices $C.D$ and $D.C$.

3. If $L = \begin{pmatrix} 1 & -2 \\ -3 & 4 \end{pmatrix}$, $M = \begin{pmatrix} 5 & 0 \\ -1 & 3 \end{pmatrix}$, form the matrices $L.M$ and $M.L$.

4. If $P = \begin{pmatrix} 1 & 0 \\ 0 & -1 \end{pmatrix}$, form the matrix $P.P$ or P^2. What is the geometrical significance of this? Can you write down the matrices P^{10}, P^{11} without working them out?

5. Write down the answers to the following straight away. If you cannot do so, work them out.

(a) $\begin{pmatrix} -1 & 0 \\ 0 & 1 \end{pmatrix}^2$

(b) $\begin{pmatrix} -1 & 0 \\ 0 & 1 \end{pmatrix} . \begin{pmatrix} -1 & 0 \\ 0 & -1 \end{pmatrix}$

(c) $\begin{pmatrix} 1 & 0 \\ 0 & -1 \end{pmatrix} . \begin{pmatrix} -1 & 0 \\ 0 & -1 \end{pmatrix}$

(d) $\begin{pmatrix} -1 & 0 \\ 0 & -1 \end{pmatrix}^3$

(e) $\begin{pmatrix} a & b \\ c & d \end{pmatrix} . \begin{pmatrix} 1 & 0 \\ 0 & 1 \end{pmatrix}$

(f) $\begin{pmatrix} a & b \\ c & d \end{pmatrix} . \begin{pmatrix} 0 & 0 \\ 0 & 0 \end{pmatrix}$

6. If $A = \begin{pmatrix} 1 & 1 \\ 2 & 2 \end{pmatrix}$, $B = \begin{pmatrix} 1 & -1 \\ -1 & 1 \end{pmatrix}$, form the matrices $A.B$ and $B.A$. What do you notice about the result?

7. Fig. 2 shows the positions of a line OP_0 after anti-clockwise rotations of 1, 2 and 3 right angles.

Show that $\begin{matrix} x_1 = -y_0 \\ y_1 = x_0 \end{matrix}$, and hence that P_0 is mapped on to or transformed into the point P_1 by the operation of the matrix $\begin{pmatrix} 0 & -1 \\ 1 & 0 \end{pmatrix}$. (Call this A.)

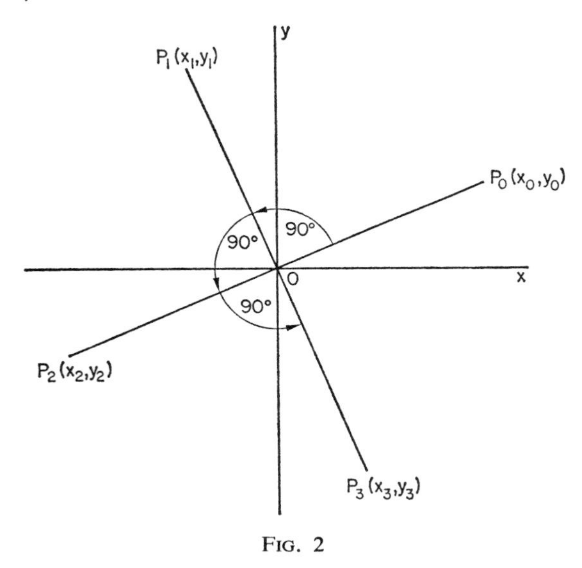

FIG. 2

Now find the matrices which:

(a) transform the point P_0 into the point P_2 (B),
(b) transform the point P_0 into the point P_3 (C),
(c) transform the point P_0 into itself (I).

Show that the matrices $\{I, A, B, C\}$ form a group. Show further that this group is isomorphic with the group of rotations of the square $\{I, \omega, \omega^2, \omega^3\}$ $(\omega^4 = 1)$.

8. P is the point (1,1) and Q is the point (2,2). Transform P and Q by means of the following matrices:

(a) $\begin{pmatrix} 1 & 0 \\ 0 & 1 \end{pmatrix}$ (b) $\begin{pmatrix} 1 & 0 \\ 0 & -1 \end{pmatrix}$ (c) $\begin{pmatrix} -1 & 0 \\ 0 & 1 \end{pmatrix}$

(d) $\begin{pmatrix} \frac{1}{\sqrt{2}} & -\frac{1}{\sqrt{2}} \\ \frac{1}{\sqrt{2}} & \frac{1}{\sqrt{2}} \end{pmatrix}$ (e) $\begin{pmatrix} 1 & -1 \\ 1 & 1 \end{pmatrix}$ (f) $\begin{pmatrix} 2 & 0 \\ 0 & 2 \end{pmatrix}$

(g) $\begin{pmatrix} 2 & 1 \\ 1 & 2 \end{pmatrix}$

Describe what has happened to the segment PQ in each case in terms of translation, reflection, rotation and enlargement.

9. If $x_3 = x_2+y_2$; $x_2 = x_1+2y_1$; $x_1 = 2x_0-y_0$;

$y_3 = x_2-y_2$ $y_2 = 2x_1-y_1$ $y_1 = x_0+2y_0$

express x_3, y_3 in terms of x_0 and y_0.

10. Construct the matrix which, operating upon the segment OP where P is the point (1,1),

(a) doubles its length,
(b) triples its length,
(c) gives its reflection in the X axis,
(d) gives its reflection in the Y axis,
(e) rotates it anti-clockwise about 0 through 180°,
(f) rotates it anti-clockwise about 0 through 90°,
(g) rotates it anti-clockwise about 0 through 45°,
(h) rotates it anti-clockwise about 0 through 90° and doubles its length.

11. What is the effect of the matrix $\begin{pmatrix} 1 & \frac{1}{2} \\ \frac{1}{2} & 1 \end{pmatrix}$ upon the square whose vertices are (0,0); (1,0); (1,1); (0,1)?

12. What is the effect of the matrix $\begin{pmatrix} 1 & 1 \\ 1 & 1 \end{pmatrix}$ upon the square whose vertices are (0,0); (1,0); (1,1); (0,1)?

MATRIX MULTIPLICATION IN ARITHMETIC

Certain types of problem in arithmetic involve exactly the same sort of multiplication as we have been doing with matrices. Consider, for example, the following problem.

In a certain coal mine, the amounts of Grade 1 and Grade 2 coal (in tons) obtained per shift from each of two seams, A and B are given by the following table:

	Grade 1	Grade 2
Seam A	4,000	2,000
Seam B	1,000	3,000

Seam A is worked 5 shifts per week and seam B is worked 4 shifts per week. Grade 1 coal sells at £9 per ton and Grade 2 coal sells at £8 per ton. Find:

1. the total amount of coal mined each week,
2. the market value of the coal mined each shift,
3. the market value of the coal mined each week.

1. The total amount of coal mined each week is obviously

$$(5\times 4{,}000+4\times 1{,}000) \text{ tons of Grade 1}$$

and $$(5\times 2{,}000+4\times 3{,}000) \text{ tons of Grade 2,}$$

or (24,000 , 22,000) tons of Grade 1 and Grade 2 respectively. Notice that this is exactly what we obtain from the matrix product

$$(5 \;,\; 4) . \begin{pmatrix} 4{,}000 & 2{,}000 \\ 1{,}000 & 3{,}000 \end{pmatrix}$$

which equals $(5\times 4{,}000+4\times 1{,}000 \;,\; 5\times 2{,}000+4\times 3{,}000)$

or (24,000 , 22,000) tons of Grade 1 and Grade 2 respectively.

2. The market value of the coal per shift, however, is:

$$\begin{matrix}\text{Seam A} & \pounds \\ \text{Seam B} & \pounds\end{matrix}\begin{pmatrix}9\times 4{,}000+8\times 2{,}000 \\ 9\times 1{,}000+8\times 3{,}000\end{pmatrix} \text{ or } \begin{pmatrix}\pounds 52{,}000 \\ \pounds 33{,}000\end{pmatrix}\begin{matrix}\text{seam A} \\ \text{seam B}\end{matrix}$$

or £85,000 altogether per shift.

This is exactly what we obtain from the matrix product

$$\begin{pmatrix}4{,}000 & 2{,}000 \\ 1{,}000 & 3{,}000\end{pmatrix}\cdot\begin{pmatrix}9 \\ 8\end{pmatrix} \text{ or } \begin{pmatrix}9\times 4{,}000+8\times 2{,}000 \\ 9\times 1{,}000+8\times 3{,}000\end{pmatrix}$$

or

$$\begin{pmatrix}52{,}000 \\ 33{,}000\end{pmatrix}\begin{matrix}\text{seam A} \\ \text{seam B}\end{matrix}$$

the units being in £.

3. The market value of the coal mined per week is the market value of coal per shift × the number of shifts worked in seam A + the market value of coal per shift in seam B × the number of shifts worked in seam B. The market value in £ is:

$$(52{,}000\times 5+33{,}000\times 4) \text{ or } \pounds 392{,}000.$$

Now this is exactly what we obtain from the matrix product

$$(5\ ,\ 4)\cdot\begin{pmatrix}52{,}000 \\ 33{,}000\end{pmatrix}$$

or, starting with the initial data, the total market value of coal mined per week is equal to

$$\pounds(5\ ,\ 4)\cdot\begin{pmatrix}4{,}000 & 2{,}000 \\ 1{,}000 & 3{,}000\end{pmatrix}\cdot\begin{pmatrix}9 \\ 8\end{pmatrix}$$

Since matrix multiplication is associative, it does not matter whether we find the product of the first two and then multiply by the *column matrix* $\begin{pmatrix}9 \\ 8\end{pmatrix}$, or multiply the *row matrix* $(5\ ,\ 4)$ by the product of the last two matrices. In this triple product a mathematician would say that $\begin{pmatrix}4{,}000 & 2{,}000 \\ 1{,}000 & 3{,}000\end{pmatrix}$ is *pre-multiplied* by $(5\ ,\ 4)$ and *post-multiplied* by $\begin{pmatrix}9 \\ 8\end{pmatrix}$.

Exercise 7(*c*)

1. Two types of food, 1 and 2, have a vitamin content in units per lb given by the following table:

	Vit. A	Vit. B
Food 1	3	7
Food 2	2	9

Express the vitamin content of 5 lb of food 1 and 6 lb of food 2 as a matrix product and evaluate it. If food 1 costs 3/- per lb and food 2 costs 3/6 per lb, express the cost of 5 lb, 6 lb of foods 1, 2 respectively as a matrix product and evaluate it.

2. A motor corporation has two types of factories each producing saloons and vans. The weekly production figures at each type of factory are as follows:

	Factory A	Factory B
Saloons	20	30
Vans	40	10

The corporation has 5 factories A and 7 factories B. Saloons and vans sell at £500, £400 respectively. Express in matrix form and hence evaluate:

(i) The total weekly production of saloons and vans.
(ii) The total market value of vehicles produced each week.

3. A builder develops a site by building 9 houses and 6 bungalows. On the average one house requires 1,600 units of materials and 2,000 hours of labour; one bungalow requires 1,500 units of materials and 1,800 hours of labour. Labour costs 10/- per hour and each unit of material costs, on the average, 20/-. Express in matrix form and hence evaluate:

(i) The total materials and labour used in completing the site.
(ii) The cost of building a house and a bungalow.
(iii) The total cost of developing the site.

4. Two television companies, TV1 and TV2, both televise documentary programmes and variety programmes. TV1 has two transmitting stations and TV2 has three transmitting stations. All stations transmit different programmes. On average the TV1 stations broadcast 1 hour of documentary and 3 hours of variety programmes each day, whereas each TV2 station broadcasts 2 hours of documentary and $1\frac{1}{2}$ hours of variety programmes each day. The transmission of documentary and variety programmes costs approximately £50 and £200 per hour respectively. Express in matrix form and hence evaluate:

(i) The daily cost of transmission from each TV1 and each TV2 station.

(ii) The total number of hours daily which are devoted to documentary and to variety programmes by both corporations.

(iii) The total daily cost of transmission incurred by both corporations.

5. The total cost of manufacturing three types of motor car is given by the following table:

	Labour (hrs)	*Materials (units)*	*Sub-contracted work (units)*
Car A	40	100	50
Car B	80	150	80
Car C	100	250	100

Labour costs £2 per hour, units of material cost 10/- each and units of sub-contracted work cost £1 per unit. Find the total cost of manufacturing 3,000, 2,000 and 1,000 vehicles of type A, B and C respectively.
(Express the cost as a triple product of a three element row matrix, a 3×3 matrix and a three element column matrix and perform the multiplication according to the same rules you used for 2×2 matrices.)

TRIPLE MATRIX PRODUCTS IN GEOMETRY

(omit at first reading)

For some reasons it is convenient to write the equations of conic sections in matrix form. The equation of the circle centre (0 , 0), radius r, is given by

$$x^2+y^2 = r^2$$

or

$$(x\ ,\ y)\,.\begin{pmatrix}x\\y\end{pmatrix} = r^2.$$

In general, the central conic $ax^2+2hxy+by^2 = c$ may be written as the triple product $(x\ ,\ y)\,.\begin{pmatrix}a & h\\h & b\end{pmatrix}.\begin{pmatrix}x\\y\end{pmatrix} = c.$

Exercise 7(*d*)

1. If $(x\ ,\ y)\,.\begin{pmatrix}2 & -2\frac{1}{2}\\-2\frac{1}{2} & 2\end{pmatrix}.\begin{pmatrix}x\\y\end{pmatrix} = 0$, express x in terms of y.

2. For what values of a does the triple matrix product $(a\ ,\ 1)\,.\begin{pmatrix}1 & -2\\-2 & 3\end{pmatrix}.\begin{pmatrix}a\\1\end{pmatrix}$ vanish?

3. Sketch the graphs of the following:

(a) $(x\ ,\ y)\,.\begin{pmatrix}1 & 0\\0 & 4\end{pmatrix}.\begin{pmatrix}x\\y\end{pmatrix} = 4$

(b) $(x\ ,\ y)\,.\begin{pmatrix}1 & 0\\0 & -1\end{pmatrix}.\begin{pmatrix}x\\y\end{pmatrix} = 1$

4. Shade the regions in which x, y are satisfied by:

(a) $(x\ ,\ y)\,.\begin{pmatrix}x\\y\end{pmatrix} \leqq 1$

(b) $(x\ ,\ y)\,.\begin{pmatrix}4 & 0\\0 & 9\end{pmatrix}.\begin{pmatrix}x\\y\end{pmatrix} \leqq 36$

5. Apply the transformation $\begin{pmatrix} X \\ Y \end{pmatrix} = \begin{pmatrix} 1 & 0 \\ 0 & k \end{pmatrix} \cdot \begin{pmatrix} x \\ y \end{pmatrix}$ $(0 < k < 1)$ to the unit circle $x^2 + y^2 = 1$ and show that the resulting figure is an ellipse with a major semi-axis one unit long.

If you have done some coordinate geometry, find for what values of k (a) the ellipse has half the area of the circle,

(b) the ellipse has an eccentricity of $\frac{1}{2}$.

EQUALITY OF MATRICES

We have seen that a matrix is an operator. It has no numerical value. When we say that two matrices are *equal* we mean that the effects of their operations upon some point (x, y) are identical. But, having done Ex. 7(a) and 7(b), we have also probably realized that there is one, and only one, matrix which will produce a given transformation in, say, the point (x, y). For example, this point is mapped on to its reflection in the x axis only by the matrix $\begin{pmatrix} 1 & 0 \\ 0 & -1 \end{pmatrix}$. We say, therefore, that two matrices are equal if (1) they are of the same order,

(2) their corresponding elements are equal.

Example. (i) $\begin{pmatrix} 1 & 2 \\ 3 & 4 \end{pmatrix} = \begin{pmatrix} 1 & 2 \\ 3 & 4 \end{pmatrix}$

(ii) $\begin{pmatrix} 1 & 2 \\ 3 & 4 \end{pmatrix} \neq \begin{pmatrix} 1 & 2 & 0 \\ 3 & 4 & 0 \\ 0 & 0 & 0 \end{pmatrix}$

(iii) $\begin{pmatrix} 1 & 2 \\ 3 & 4 \end{pmatrix} \neq \begin{pmatrix} 1 & 3 \\ 2 & 4 \end{pmatrix}$

(iv) If $\begin{pmatrix} x+2 \\ y-4 \end{pmatrix} = \begin{pmatrix} 3 \\ 1 \end{pmatrix}$

then $x+2 = 3$ and $y-4 = 1$

i.e. $x = 1$ and $y = 5$.

(v) If $\begin{pmatrix} x+y & x-y \\ p+q & p-q \end{pmatrix} = \begin{pmatrix} 3 & 1 \\ 7 & 5 \end{pmatrix}$

then $x+y = 3 \quad p+q = 7$

$x-y = 1 \quad p-q = 5$

i.e. $x = 2,\ y = 1;\ p = 6,\ q = 1.$

Exercise 7(*e*)

1. Find the values of x, y, z, such that $\begin{pmatrix} x-3 \\ y+4 \\ z-2 \end{pmatrix} = \begin{pmatrix} 2 \\ 5 \\ 1 \end{pmatrix}$

2. Find the values x, y, a, b, such that

$$\begin{pmatrix} x-2y & 2x-y \\ a+b & a-b \end{pmatrix} = \begin{pmatrix} 0 & 3 \\ 7 & 1 \end{pmatrix}$$

3. Find the values of x, y, z, w, such that

$$\begin{pmatrix} x+y & 2x-y \\ y+z & w-x \end{pmatrix} = \begin{pmatrix} 3 & 0 \\ 5 & 3 \end{pmatrix}$$

4. Find the values of x, y, such that $(x^2-y^2\ ,\ x+y) = (7\ ,\ 1)$.

5. Find the values of x, y, z, w, such that

$$\begin{pmatrix} x^2-x & y^2 \\ z^2+4 & w^2 \end{pmatrix} = \begin{pmatrix} 2x+4 & y+2 \\ 4z & w \end{pmatrix}$$

ADDITION OF MATRICES

Consider the column matrices $\begin{pmatrix} a \\ b \end{pmatrix}$, $\begin{pmatrix} c \\ d \end{pmatrix}$. What do we mean by "adding" these together? Earlier in the chapter we used column matrices like these to represent points. What do we mean by "adding" points together?

Well, if we think of $\begin{pmatrix} a \\ b \end{pmatrix}$ as being a displacement a along $\overrightarrow{OX}$, followed by a displacement b in the direction $\overrightarrow{OY}$, then, since these two displacements are equivalent to the direct displacement or *vector* $\overrightarrow{OP}$, $\begin{pmatrix} a \\ b \end{pmatrix}$ is a matrix form of the vector $\overrightarrow{OP}$. For this reason we sometimes call it a *column vector*.

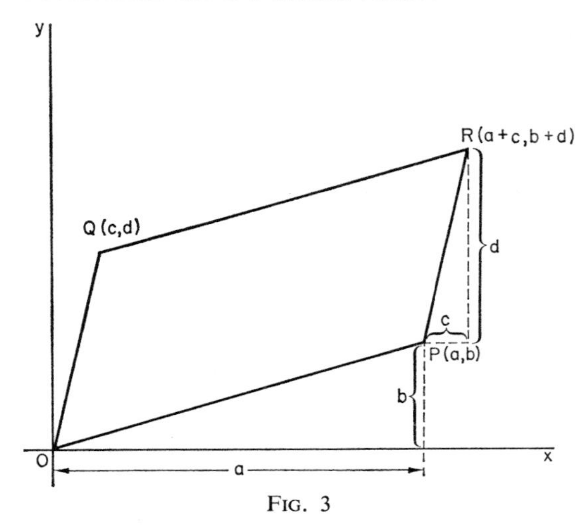

FIG. 3

If we now add the column vector $\begin{pmatrix} c \\ d \end{pmatrix}$, i.e. if we make a further displacement c in the direction $\overrightarrow{OX}$, followed by a displacement d in the direction $\overrightarrow{OY}$, we arrive at the point R which is represented by the column vector $\begin{pmatrix} a+c \\ b+d \end{pmatrix}$. Again, these further displacements are equivalent to the further direct displacement $\overrightarrow{PR}$. Now a displacement $\overrightarrow{OP}$ followed by a displacement $\overrightarrow{PR}$ is equivalent to a direct displacement $\overrightarrow{OR}$, so we may write:

$$\overrightarrow{OP}+\overrightarrow{PR} = \overrightarrow{OR}$$

Or, in terms of column matrices (or column vectors):

$$\begin{pmatrix} a \\ b \end{pmatrix}+\begin{pmatrix} c \\ d \end{pmatrix} = \begin{pmatrix} a+c \\ b+d \end{pmatrix} \qquad \text{(III)}$$

Notice in Fig. 3 that $\vec{OQ}$ is also represented by $\begin{pmatrix} c \\ d \end{pmatrix}$, hence $\vec{OQ} = \vec{PR}$. Also, $OPRQ$ is a parallelogram, so what we are really doing is to add our matrices in such a way that they obey the parallelogram law of addition. In equation III we have formed the sum of two matrices by adding together corresponding elements. We now take this as a general law of addition of any two matrices of equal order.

Example. (i) $\begin{pmatrix} 1 & 2 \\ 4 & -3 \end{pmatrix} + \begin{pmatrix} 0 & 6 \\ 3 & 2 \end{pmatrix} = \begin{pmatrix} 1 & 8 \\ 7 & -1 \end{pmatrix}$

(ii) $\begin{pmatrix} 7 & 5 \\ 0 & 3 \end{pmatrix} - \begin{pmatrix} 2 & 4 \\ 1 & 3 \end{pmatrix} = \begin{pmatrix} 5 & 1 \\ -1 & 0 \end{pmatrix}$

(iii) $\begin{pmatrix} 2 \\ 7 \\ 5 \end{pmatrix} + \begin{pmatrix} 6 \\ 1 \\ 4 \end{pmatrix} = \begin{pmatrix} 8 \\ 8 \\ 9 \end{pmatrix}$

(iv) $(1, 2, 3) + (2, 0, 4) = (3, 2, 7)$

(v) $2\begin{pmatrix} 1 & 2 \\ 3 & 4 \end{pmatrix} = \begin{pmatrix} 1 & 2 \\ 3 & 4 \end{pmatrix} + \begin{pmatrix} 1 & 2 \\ 3 & 4 \end{pmatrix} = \begin{pmatrix} 2 & 4 \\ 6 & 8 \end{pmatrix}$

(vi) $\lambda\begin{pmatrix} 1 & 2 \\ 3 & 4 \end{pmatrix} = \begin{pmatrix} 1 & 2 \\ 3 & 4 \end{pmatrix} \ldots \underset{\text{times}}{\lambda} = \begin{pmatrix} \lambda & 2\lambda \\ 3\lambda & 4\lambda \end{pmatrix}$

Exercise 7(f)

1. If $X = \begin{pmatrix} 1 & 3 \\ 4 & 2 \end{pmatrix}$, $Y = \begin{pmatrix} 4 & -2 \\ 0 & 5 \end{pmatrix}$, $Z = \begin{pmatrix} 3 & -1 \\ 1 & 3 \end{pmatrix}$, find the following matrices:

(a) $X+Y$ (b) $X+Z$ (c) $Y+Z$
(d) $2X$ (e) $3Y$ (f) $4Z$
(g) $X-Y$ (h) $Y-2Z$ (i) $2X+3Y-4Z$

2. Find the matrix A satisfying the following matrix equations:

(a) $A + \begin{pmatrix} 2 & 4 \\ 3 & 5 \end{pmatrix} = \begin{pmatrix} 7 & 2 \\ 0 & 1 \end{pmatrix}$

(b) $2A - \begin{pmatrix} 1 & 4 \\ 3 & 7 \end{pmatrix} = \begin{pmatrix} 3 & 2 \\ 5 & 1 \end{pmatrix}$

(c) $3A+\begin{pmatrix}1 & 2 & 3\\ 4 & 7 & 9\\ 0 & 2 & 1\end{pmatrix} = \begin{pmatrix}10 & 8 & 0\\ -2 & 4 & 3\\ 3 & -1 & 16\end{pmatrix}$

(d) $A+\begin{pmatrix}1\\ 2\\ 3\end{pmatrix} = 2\begin{pmatrix}3\\ 5\\ 4\end{pmatrix} - 3\begin{pmatrix}1\\ 0\\ 2\end{pmatrix}$

(e) $A+(-2, 4, 7) = (2, 5, -4)-(1, 0, 2)$

3. If X is the column vector $\begin{pmatrix}1\\ 3\end{pmatrix}$, Y is the column vector $\begin{pmatrix}3\\ 1\end{pmatrix}$, represent *graphically* the following column vectors. (The axes are rectangular.)

(a) $X+Y$ (b) $X-Y$ (c) $\frac{1}{2}(X+Y)$
(d) $2X+Y$ (e) $\frac{1}{3}(2X+Y)$ (f) $\frac{1}{3}(X+2Y)$
(g) $3X-Y$ (h) $X-3Y$ (i) $\frac{1}{5}(2X+3Y)$

What do you notice about the results (c), (e), (f), (i)?

4. Repeat Question 3 using axes yox where $\angle yox = 60°$. Show that the process of column vector addition is independent of the axes chosen and that the results (c), (e), (f), (i), still have the same geometrical significance.

5. P, Q, R, are the column vectors $\begin{pmatrix}4\\ 1\end{pmatrix}$, $\begin{pmatrix}-1\\ 4\end{pmatrix}$, $\begin{pmatrix}-6\\ 7\end{pmatrix}$.

Represent these graphically on rectangular axes by vectors $\overrightarrow{OP}$, $\overrightarrow{OQ}$, $\overrightarrow{OR}$.

(a) Show that Q is the midpoint of the line PR.
(b) Show that $Q-P = R-Q$.
(c) Show that $Q = \frac{1}{2}(P+R)$.
(d) Represent the column vector $\frac{1}{5}(2P+3Q)$ by $\overrightarrow{OS}$. Show that S divides PQ in the ratio 3:2.
(e) Show that $\overrightarrow{OP}$ and $\overrightarrow{OQ}$ are perpendicular.
(f) Form the matrix products $(-1, 4)\cdot\begin{pmatrix}4\\ 1\end{pmatrix}$; $(4, 1)\cdot\begin{pmatrix}-1\\ 4\end{pmatrix}$.

What do you notice about the results (a), (b), (c), and (e), (f). Can you generalize (d)?

THE INVERSE MATRIX

In the last chapter we saw that in any group of operations the effect of any one of the operations can always be "undone" by itself or by another operation in the same group. Furthermore, in each case there was only one operation which would do this. In this chapter we have seen that any given 2×2 matrix (call it A) may act as an operator. We now look at the problem of finding if there is another 2×2 matrix (say B) which will cancel its effect. In terms of the identity of unit matrix (which leaves things unchanged) $\begin{pmatrix} 1 & 0 \\ 0 & 1 \end{pmatrix}$, (call it I), then, given A, we wish to find B such that $B.A = I$. To use a notation already familiar to us we denote B, where it exists, by A^{-1}, so that $A^{-1}.A = I$. A practical need to find such a matrix arises in the solution of simultaneous equations.

Suppose we have
$$ax+by = p$$
$$cx+dy = q$$

or, in matrix form,
$$\begin{pmatrix} a & b \\ c & d \end{pmatrix} \cdot \begin{pmatrix} x \\ y \end{pmatrix} = \begin{pmatrix} p \\ q \end{pmatrix}.$$

Then, writing $\begin{pmatrix} a & b \\ c & d \end{pmatrix}$ as A, and its *inverse*, or the matrix which "undoes" it, as A^{-1}, we have:

$$A\begin{pmatrix} x \\ y \end{pmatrix} = \begin{pmatrix} p \\ q \end{pmatrix}$$

$$\therefore\ A^{-1} . A\begin{pmatrix} x \\ y \end{pmatrix} = A^{-1} . \begin{pmatrix} p \\ q \end{pmatrix}$$

$$\therefore\ I . \begin{pmatrix} x \\ y \end{pmatrix} = A^{-1} . \begin{pmatrix} p \\ q \end{pmatrix}$$

or
$$\begin{pmatrix} x \\ y \end{pmatrix} = A^{-1} . \begin{pmatrix} p \\ q \end{pmatrix}$$

Thus, if we can find A^{-1}, we have the solution.

Exercise 7(g)

1. Before reading the next paragraph see if you can find A^{-} for yourself. Using this as a general rule, write down straight away the inverses of the following matrices:

(a) $\begin{pmatrix} 2 & 3 \\ 1 & 2 \end{pmatrix}$ (b) $\begin{pmatrix} 3 & 1 \\ 2 & 1 \end{pmatrix}$ (c) $\begin{pmatrix} -1 & 0 \\ 0 & -1 \end{pmatrix}$

(d) $\begin{pmatrix} 1 & 0 \\ 0 & -1 \end{pmatrix}$ (e) $\begin{pmatrix} -1 & 0 \\ 0 & 1 \end{pmatrix}$ (f) $\begin{pmatrix} 1 & 0 \\ 0 & 1 \end{pmatrix}$

(g) $\begin{pmatrix} 2 & 3 \\ 4 & -1 \end{pmatrix}$ (h) $\begin{pmatrix} 4 & -2 \\ 3 & 1 \end{pmatrix}$

What do you notice about the results (c), (d), (e), (f)? Could you have written down these results without knowing the inverse of A?

2. Can you find the inverses of the following matrices?

(a) $\begin{pmatrix} 2 & 4 \\ 1 & 2 \end{pmatrix}$ (b) $\begin{pmatrix} 3 & 1 \\ 6 & 2 \end{pmatrix}$ (c) $\begin{pmatrix} 1 & -1 \\ -1 & 1 \end{pmatrix}$

Can you give a general rule for cases in which A does not possess an inverse? What is the algebraical and geometrical significance of such cases?

If you were able to do Exercise 7(g) correctly you need not read the following paragraph. If you had difficulty, however, you should have proceeded as follows:

$$ax+by = p \tag{1}$$
$$cx+dy = q \tag{2}$$

Multiply example 1 by d, example 2 by b, and subtract.

Thus
$$dax+dby = dp$$
$$bcx+bdy = bq$$

$$\therefore \; x(da-bc) = dp-bq$$

$$\therefore \; x = \frac{dp-bq}{da-bc}$$

Similarly $$y = \frac{-pc+qa}{da-bc}$$

or $$\begin{pmatrix} x \\ y \end{pmatrix} = \frac{1}{ad-bc}\begin{pmatrix} d & -b \\ -c & a \end{pmatrix} \cdot \begin{pmatrix} p \\ q \end{pmatrix}$$

Thus, the inverse of A, i.e. of $\begin{pmatrix} a & b \\ c & d \end{pmatrix}$, is $\dfrac{1}{ad-bc}\begin{pmatrix} d & -b \\ -c & a \end{pmatrix}$.

In words; we make b, c negative, interchange a and d, and divide by the cross-product of the elements of matrix A.

(The cross-product of the array $\begin{matrix} a & b \\ c & d \end{matrix}$ is $ad-bc$. As an example, the cross-product of the array $\begin{matrix} 2 & 3 \\ 3 & -1 \end{matrix}$ is $2 \times -1 - 3 \times 3$, i.e. -11.)

Later you will learn that $ad-bc$ is called the *determinant* of the matrix $\begin{pmatrix} a & b \\ c & d \end{pmatrix}$ and is written $\begin{vmatrix} a & b \\ c & d \end{vmatrix}$. Furthermore, when its value is zero, we are unable to find finite values of x, y satisfying the original equations. A matrix whose determinant is zero is called a *singular* matrix.

Examples. (i) Solve the equations $\begin{aligned} 2x+3y &= 7 \\ 3x-y &= 5 \end{aligned}$

$$\therefore \begin{pmatrix} 2 & 3 \\ 3 & -1 \end{pmatrix} \cdot \begin{pmatrix} x \\ y \end{pmatrix} = \begin{pmatrix} 7 \\ 5 \end{pmatrix}$$

The inverse of $\begin{pmatrix} 2 & 3 \\ 3 & -1 \end{pmatrix}$ is $-\dfrac{1}{11}\begin{pmatrix} -1 & -3 \\ -3 & 2 \end{pmatrix}$

$$\therefore \begin{pmatrix} x \\ y \end{pmatrix} = -\frac{1}{11}\begin{pmatrix} -1 & -3 \\ -3 & 2 \end{pmatrix} \cdot \begin{pmatrix} 7 \\ 5 \end{pmatrix}$$

$$= -\frac{1}{11}\begin{pmatrix} -22 \\ -11 \end{pmatrix}$$

$$= \begin{pmatrix} 2 \\ 1 \end{pmatrix}$$

$$\therefore x = 2, \ y = 1.$$

(ii) Solve the equations $2x+4y = 7$ (1)

$x+2y = 3$ (2)

Here
$$\begin{pmatrix} 2 & 4 \\ 1 & 2 \end{pmatrix} \cdot \begin{pmatrix} x \\ y \end{pmatrix} = \begin{pmatrix} 7 \\ 3 \end{pmatrix}$$

The inverse of $\begin{pmatrix} 2 & 4 \\ 1 & 2 \end{pmatrix}$ is $\frac{1}{0}\begin{pmatrix} 2 & -4 \\ -1 & 2 \end{pmatrix}$.

In this case we cannot determine x and y. The reason for this becomes obvious if we draw the graphs of 1 and 2. They are parallel straight lines and hence do not intersect. There are therefore no finite values of x, y which satisfy these equations.

Exercise 7(*h*)

1. By finding an inverse matrix in each case, solve the following pairs of simultaneous equations for values of x and y:

(a) $\begin{aligned} 2x+y &= 4 \\ x+y &= 3 \end{aligned}$ (b) $\begin{aligned} 3x+y &= 9 \\ 5x+2y &= 16 \end{aligned}$ (c) $\begin{aligned} 4x+3y &= 5 \\ x+y &= 2 \end{aligned}$

(d) $\begin{aligned} 2x+y &= 4 \\ x+2y &= 5 \end{aligned}$ (e) $\begin{aligned} 3x+2y &= 1 \\ 2x-3y &= 5 \end{aligned}$ (f) $\begin{aligned} 4x-y &= 9 \\ x-2y &= 4 \end{aligned}$

(g) $x+3y = 2x+y-1 = 6$

(h) $\begin{aligned} 2x-y &= p+3q \\ 3x+y &= 2(2p+q) \end{aligned}$

2. Find the inverse of the matrix $\begin{pmatrix} 4 & -2 \\ 3 & 1 \end{pmatrix}$. Use this inverse matrix to solve the following simultaneous equations:

(a) $\begin{aligned} 4x-2y &= p \\ 3x+y &= q \end{aligned}$ (b) $\begin{aligned} 2x-y &= 6 \\ 3x+y &= 14 \end{aligned}$ (c) $\begin{aligned} y &= 2x \\ y &= 5-3x \end{aligned}$

(d) $4x-2y = 5 = 6x+2y$

3. $A = \begin{pmatrix} 5 & 3 \\ 3 & 2 \end{pmatrix}$, $B = \begin{pmatrix} 2 & 1 \\ 1 & 1 \end{pmatrix}$.

Find (a) A^{-1} (b) B^{-1} (c) AB
(d) BA (e) $(AB)^{-1}$ (f) $B^{-1}.A^{-1}$
(g) $A^{-1}.B^{-1}$ (h) $A^{-1}.A$

What can you say about A, B and their products?

4. $A = \begin{pmatrix} 2 & 3 \\ 1 & 2 \end{pmatrix}$, $B = \begin{pmatrix} 1 & 1 \\ 1 & 2 \end{pmatrix}$.

Find (a) A^{-1} (b) B^{-1} (c) $A.B$
(d) $B.A$ (e) $(AB)^{-1}$ (f) $B^{-1}.A^{-1}$
(g) $(BA)^{-1}$ (h) $A^{-1}.B^{-1}$

What do you notice about these results?

5. Suppose $A = \begin{pmatrix} 2 & 1 \\ 1 & 2 \end{pmatrix}$, $B = \begin{pmatrix} 3 & 2 \\ 2 & 3 \end{pmatrix}$. Notice that in both cases the sum of any row or column is a constant.

[They are very simple examples of what we call *magic matrices*, although in a true magic matrix the diagonal sums are also constant.]

Investigate the following matrices:

(a) $A.B$ (b) A^{-1} (c) B^{-1} (d) $(A.B)^{-1}$

If A has magic number 3 (its row and column constant) what is the magic number of A^{-1}? Is the same result true of B and B^{-1}, $A.B$ and $(A.B)^{-1}$?
Try to show that, in general, the product of any 2×2 magic matrices, $\begin{pmatrix} a & h \\ h & a \end{pmatrix}$, $\begin{pmatrix} b & k \\ k & b \end{pmatrix}$, is magic and that the inverses of such matrices are also magic.

6. If we are given a matrix $A = \begin{pmatrix} a & b \\ c & d \end{pmatrix}$ and we interchange its rows and columns, we obtain what is called the *transposed matrix* $\tilde{A}$ or $\begin{pmatrix} a & c \\ b & d \end{pmatrix}$. [Later we see that there are close connections between $\tilde{A}$ and A^{-1}. For certain important matrices called *orthogonal* matrices we actually have $\tilde{A} = A^{-1}$.]

Suppose now that $P = \begin{pmatrix} 2 & 3 \\ 1 & 4 \end{pmatrix}$, $Q = \begin{pmatrix} 1 & 4 \\ 2 & 3 \end{pmatrix}$.

Form the matrices (a) $\tilde{P}$ (b) $\tilde{Q}$ (c) $P.Q$
(d) $\widetilde{P.Q}$ (e) $\tilde{P}.\tilde{Q}$ (f) $\tilde{Q}.\tilde{P}$

What do you notice about these results?

7. Make the substitutions $\begin{matrix} x = aX+bY \\ y = cX+dY \end{matrix}$ in the equation $x^2+y^2 = 1$.

Show that if $\begin{pmatrix} x \\ y \end{pmatrix} = A\begin{pmatrix} X \\ Y \end{pmatrix}$ where $A = \begin{pmatrix} a & b \\ c & d \end{pmatrix}$

then $$(x,y) = (X, Y).\tilde{A}.$$

Hence show that the original substitutions can be made by writing $x^2+y^2 = 1$, or $(x,y) . \begin{pmatrix} x \\ y \end{pmatrix} = 1$, and the resulting equation in X, Y as $(X, Y) . \tilde{A} . A . \begin{pmatrix} X \\ Y \end{pmatrix} = 1$.

Use this result to make the substitution $\begin{matrix} x = 4X-2Y \\ y = 3X+Y \end{matrix}$ in the equation $x^2+y^2 = 1$.

Now make the substitution $\begin{pmatrix} x \\ y \end{pmatrix} = \begin{pmatrix} \frac{1}{\sqrt{2}} & -\frac{1}{\sqrt{2}} \\ \frac{1}{\sqrt{2}} & \frac{1}{\sqrt{2}} \end{pmatrix} . \begin{pmatrix} X \\ Y \end{pmatrix}$ in the equation $x^2+y^2 = 1$.

Can you explain this result?

8. Letters of the alphabet correspond to the numbers 1–26 as follows:

A	B	C	D	E	F	G	H	I
7	16	24	14	1	22	25	17	12
J	K	L	M	N	O	P	Q	R
13	18	6	20	4	3	23	26	10
S	T	U	V	W	X	Y	Z	
11	2	9	5	15	19	8	21	

Words are coded for secret purposes by operating on them by means of the matrix $\begin{pmatrix} 2 & 1 \\ 1 & 1 \end{pmatrix}$.

Thus, for example, VETO, or 5123, is written as $\begin{pmatrix} 5 & 1 \\ 2 & 3 \end{pmatrix}$. Then $\begin{pmatrix} 2 & 1 \\ 1 & 1 \end{pmatrix} \cdot \begin{pmatrix} 5 & 1 \\ 2 & 3 \end{pmatrix} = \begin{pmatrix} 12 & 5 \\ 7 & 4 \end{pmatrix}$, or IVAN, so that IVAN is the code word for VETO. The code word is later decoded by the inverse matrix of the decoding matrix. Thus:

$$\begin{pmatrix} 1 & -1 \\ -1 & 2 \end{pmatrix} \cdot \begin{pmatrix} 12 & 5 \\ 7 & 4 \end{pmatrix} = \begin{pmatrix} 5 & 1 \\ 2 & 3 \end{pmatrix}, \text{ or VETO.}$$

If the coder makes a mistake and postmultiplies by the coding matrix we have:

$$\begin{pmatrix} 5 & 1 \\ 2 & 3 \end{pmatrix} \cdot \begin{pmatrix} 2 & 1 \\ 1 & 1 \end{pmatrix} = \begin{pmatrix} 11 & 6 \\ 7 & 5 \end{pmatrix}, \text{ or SLAV.}$$

If the decoder now decodes this correctly, what word does he obtain?

If numbers greater than 26 arise, we simply keep subtracting 26 until we obtain a number in the range 1–26. Thus, to code the word TINY, or $\begin{pmatrix} 2 & 12 \\ 4 & 8 \end{pmatrix}$, we have:

$$\begin{pmatrix} 2 & 1 \\ 1 & 1 \end{pmatrix} \cdot \begin{pmatrix} 2 & 12 \\ 4 & 8 \end{pmatrix} = \begin{pmatrix} 8 & 32 \\ 6 & 20 \end{pmatrix} = \begin{pmatrix} 8 & 6 \\ 6 & 20 \end{pmatrix}$$

or YLLM.

Find code words for the following:

(a) ETON (b) TINS (c) LANE (d) SOOT

Now *decode* the following words:

(e) QFRX (f) HCRF (g) TJWI (h) YVPD

(With apologies to Dr. Matthews and "Yogi Bear".)

8
VECTORS
COLUMN VECTORS AND ROW VECTORS

IN Fig. 1 we have a series of points, A, B, C, D, E, F, G, H, I, J, whose positions relative to each other may be given by the distances which we must travel (a) parallel to $\overrightarrow{OX}$, (b) parallel to $\overrightarrow{OY}$, in order to get from any one point to another. Thus, to get from 0 to A, i.e. to make the displacement represented by the vector $\overrightarrow{OA}$, we may travel 1 unit in the direction $\overrightarrow{OX}$ and then 3 units in a direction parallel to $\overrightarrow{OY}$. Briefly, we may represent these as either a *column vector* $\begin{pmatrix}1\\3\end{pmatrix}$ (see Chapter 7) or as a *row vector* $(1,3)$. To make a further displacement $\overrightarrow{AB}$ we must travel 3 units in a direction parallel to $\overrightarrow{OX}$ and then 1 unit in a direction parallel to $\overrightarrow{OY}$, i.e. the vector $\overrightarrow{AB}$ may be written as the row vector $(3,1)$. Notice that $\overrightarrow{OC}$ may also be written as the row vector $(3,1)$. In other words *the vectors* $\overrightarrow{AB}$ *and* $\overrightarrow{OC}$ *which are parallel, equal in length, and equal in sense, are represented by the same row vector* $(3,1)$. *They are therefore equivalent vectors.* We can, in fact, say that they are the *same vector*, for strictly *a vector is a displacement of the whole plane*, and $\overrightarrow{AB}$, $\overrightarrow{OC}$, $\overrightarrow{FG}$, are all equivalent representations of it.

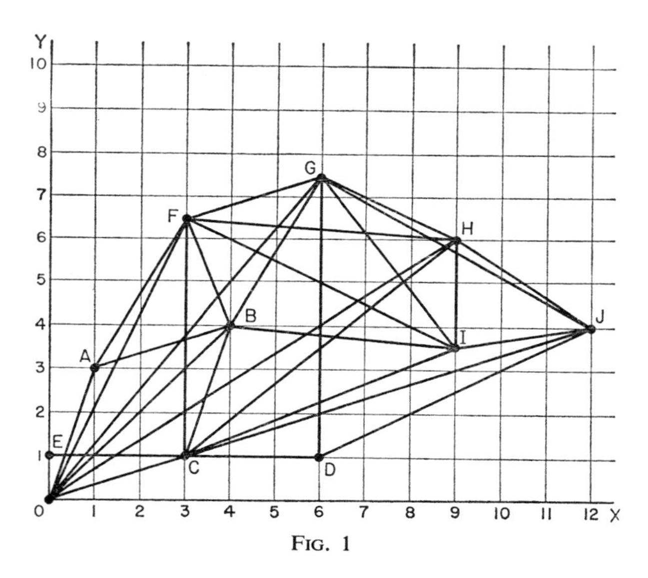

FIG. 1

VECTOR ADDITION

Now let us see what happens if a displacement $\vec{OA}$ is followed by a displacement $\vec{AB}$. Clearly, this is equivalent to the direct displacement $\vec{OB}$, i.e.

$$\vec{OA}+\vec{AB} = \vec{OB}.$$

Writing this in the form of row matrices or row vectors, and using the matrix law of addition defined in Chapter 7, we have:

$$(1,3)+(3,1) = (4,4)$$

So the matrix law of addition implies the triangle or parallelogram law of addition of vectors.

Later in this chapter we shall represent vectors such as $\vec{OA}$ by single small letters in heavy type, so we may write $\vec{OA} = \vec{CB} = \mathbf{a}$; $\vec{AB} = \vec{OC} = \mathbf{b}$; $\vec{OB} = \mathbf{c}$.

Then since $$\vec{OA}+\vec{AB} = \vec{OB}$$

and $$\vec{OC}+\vec{CB} = \vec{OB}$$

we have $$\mathbf{a}+\mathbf{b} = \mathbf{c} = \mathbf{b}+\mathbf{a}$$

i.e. *the addition of vectors is a commutative operation.* This is to be expected since the matrix law of addition is also a commutative operation, but the fact needs to be emphasized.

Now look at the vectors $\vec{CD}$ and $\vec{CE}$. $\vec{CD}$ is obviously (3,0). To get from C to E we travel *backwards* 3 *units* or *forwards* -3 units, i.e.

$$\vec{CE} = (-3,0)$$

$$\vec{CD}+\vec{CE} = (3,0)+(-3,0) = (0,0)$$

A vector (0,0) *of zero magnitude is called a null vector.*

But $$\vec{CD} = \vec{EC} = (3,0) = -(-3,0)$$

$$\therefore\ \vec{EC} = -\vec{CE}$$

i.e. *the vectors* **a** *and* $-$**a** *are parallel, equal in magnitude bu opposite in sense.*

Next we investigate what happens if we multiply a vector by a pure number (or scalar quantity). Consider the vectors $\vec{OC}$ and $\vec{OJ}$. $\vec{OC} = (3,1)$ and $\vec{OJ} = (12,4)$ or $4(3,1)$ or $4.\vec{OC}$, i.e. if we multiply the vector $\vec{OC}$ by 4 we obtain a vector which has the same direction and sense but is of four times the magnitude. In this case $\vec{OC}$ lies along the line $\vec{OJ}$, but this does not necessarily happen when we multiply by a scalar. For example, $2\frac{1}{2}\vec{OE} = 2\frac{1}{2}(0,1) = (0,2\frac{1}{2}) = \vec{IH}$. Thus, if we multiply $\vec{OE}$ by the number $2\frac{1}{2}$, we obtain a vector which is parallel, equal in sense, and two and a half times the magnitude. In general, *all that we can say of the vectors* $\mathbf{a}$ *and* $\lambda\mathbf{a}$ *is that they have the same direction and that their magnitudes are in the ratio of* $1:\lambda$. *If* λ *is positive they have the same sense; if negative, opposite senses.*

Finally, let us consider several successive displacements. By repeated application of the triangle law we can write $\overrightarrow{OJ}$ as $\overrightarrow{OA}+\overrightarrow{AF}+\overrightarrow{FG}+\overrightarrow{GH}+\overrightarrow{HJ}$, or as $\overrightarrow{OC}+\overrightarrow{CD}+\overrightarrow{DJ}$. For, writing as row vectors, we have:

$$\begin{aligned}\overrightarrow{OA}+\overrightarrow{AF}+\overrightarrow{FG}+\overrightarrow{GH}+\overrightarrow{HJ}&\\ &= (1,3)+(2,3\tfrac{1}{2})+(3,1)+(3,-1\tfrac{1}{2})+(3,-2)\\ &= (1+2+3+3+3,3+3\tfrac{1}{2}+1-1\tfrac{1}{2}-2)\\ &= (12,4) = \overrightarrow{OJ}\end{aligned}$$

and $$\begin{aligned}\overrightarrow{OC}+\overrightarrow{CD}+\overrightarrow{DJ} &= (3,1)+(3,0)+(6,3)\\ &= (12,4) = \overrightarrow{OJ}\end{aligned}$$

Exercise 8(*a*)

Questions 1–6 inclusive all refer to Fig. 1.

1. Express as row vectors:

(a) $\overrightarrow{OF}$ (b) $\overrightarrow{OG}$ (c) $\overrightarrow{OH}$ (d) $\overrightarrow{OD}$

(e) $\overrightarrow{BG}$ (f) $\overrightarrow{BH}$ (g) $\overrightarrow{IJ}$ (h) $\overrightarrow{IG}$

(i) $\overrightarrow{IB}$ (j) $\overrightarrow{BC}$ (k) $\overrightarrow{BA}$ (l) $\overrightarrow{GD}$

2. Express as positive vectors in the form $\overrightarrow{PQ}$:

(a) $-\overrightarrow{BD}$ (b) $-\overrightarrow{FI}$ (c) $-\overrightarrow{HE}$

3. Using only the letters of Fig. 1, state which vector or vectors are:

(a) equivalent to the vector $\overrightarrow{OC}$,

(b) equivalent to the vector $\overrightarrow{OA}$,

(c) equivalent to the vector $\overrightarrow{AF}$,

(d) parallel to the vector $\overrightarrow{BJ}$,

(e) parallel to the vector $\overrightarrow{GD}$,
(f) equal in magnitude (but not necessarily in direction) to the vector $\overrightarrow{CB}$,
(g) equal, in magnitude only, to the vector $\overrightarrow{BD}$.

4. Express as single vectors each of the following sums:

(a) $\overrightarrow{OA}+\overrightarrow{AF}$ (b) $\overrightarrow{FG}+\overrightarrow{GH}$ (c) $\overrightarrow{CD}+\overrightarrow{DH}$

(d) $\overrightarrow{BG}+\overrightarrow{GH}+\overrightarrow{HJ}$ (e) $\overrightarrow{BI}+\overrightarrow{IH}+\overrightarrow{HF}$

(f) $\overrightarrow{OA}+\overrightarrow{AF}+\overrightarrow{FH}+\overrightarrow{HI}+\overrightarrow{IJ}+\overrightarrow{JD}+\overrightarrow{DB}$

(g) $\overrightarrow{OB}-\overrightarrow{CB}$ (h) $\overrightarrow{OH}-\overrightarrow{CH}$ (i) $\overrightarrow{BC}-\overrightarrow{IC}$

(j) $\overrightarrow{IC}-\overrightarrow{BC}$ (k) $\overrightarrow{CD}+\overrightarrow{CE}$ (l) $\overrightarrow{GJ}-\overrightarrow{CJ}$

(m) $\overrightarrow{OH}+\overrightarrow{HI}-\overrightarrow{GI}-\overrightarrow{BG}$

(n) $\overrightarrow{OC}+\overrightarrow{CF}-\overrightarrow{IF}-\overrightarrow{DI}+\overrightarrow{DE}$

(o) $\overrightarrow{AF}+\overrightarrow{FB}+\overrightarrow{BA}$ (p) $\overrightarrow{IJ}+\overrightarrow{JI}$

(q) $\overrightarrow{OA}+\overrightarrow{AF}+\overrightarrow{FG}+\overrightarrow{GH}+\overrightarrow{HJ}+\overrightarrow{JD}+\overrightarrow{DC}+\overrightarrow{CO}$

5. Perform the following additions in terms of row vectors and give each result as a row vector.

(a) $\frac{1}{2}(\overrightarrow{OA}+\overrightarrow{OC})$ (b) $\frac{1}{5}(2\overrightarrow{BH}+3\overrightarrow{BI})$ (c) $\frac{1}{11}(10\overrightarrow{BF}+\overrightarrow{BC})$

(d) $\frac{1}{13}(6\overrightarrow{BG}+7\overrightarrow{BD})$ (e) $\frac{1}{2}(\overrightarrow{CB}-\overrightarrow{DC})$ (f) $\overrightarrow{OA}-2\overrightarrow{AF}+3\overrightarrow{FG}$

(g) $\frac{1}{3}(\overrightarrow{OA}+\overrightarrow{OB}+\overrightarrow{OC})$ (h) $\overrightarrow{OJ}-4\overrightarrow{OC}$

(i) Show that (a) is a vector $\overrightarrow{OK}$ where K is the midpoint of AC.

(j) Show that (b) is a vector $\overrightarrow{BL}$ where L divides HI in the ratio 3:2.

(k) Show that (c) is a vector $\overrightarrow{BM}$ where M divides FC in the ratio 1:10.

(l) Show that (d) is a vector $\overrightarrow{BN}$ where N divides GD in the ratio 7:6.

(m) Show that (e) is a vector $\overrightarrow{CP}$ where P is the midpoint of BD.

(n) Show that (g) is a vector $\overrightarrow{OQ}$ where Q is the centre of gravity of the triangle ABC (i.e. the point of intersection of its medians.)

6. In Fig. 1, $\overrightarrow{CD} = (3,0)$, $\overrightarrow{HI} = (0,2\frac{1}{2})$. If we write $\overrightarrow{HI}$ as a column vector $\begin{pmatrix} 0 \\ 2\frac{1}{2} \end{pmatrix}$ and form the *scalar* product $\overrightarrow{CD}.\overrightarrow{HI}$, then *scalar product* $\overrightarrow{CD}.\overrightarrow{HI}$ = *matrix product* $(3,0) . \begin{pmatrix} 0 \\ 2\frac{1}{2} \end{pmatrix} = 0$. Now $\overrightarrow{CD}$ and $\overrightarrow{HI}$ are *perpendicular vectors* and their scalar product is zero; it will be seen the same must be true of *any* pair of perpendicular vectors. Use this result to show that the following pairs of vectors are perpendicular.

(a) $\overrightarrow{AC}$, $\overrightarrow{OB}$ (b) $\overrightarrow{OH}$, $\overrightarrow{BD}$ (c) $\overrightarrow{OE}$, $\overrightarrow{EC}$

(d) $\overrightarrow{CD}$, $\overrightarrow{DG}$

FREE VECTORS

We now summarize the results obtained so far in a slightly different way. Although we have used axes OX, OY in Fig. 1, the properties that we have discovered are all independent of the coordinate system. (See Chapter 7, Exercise on addition of matrices.) In the following sections we shall use vectors without reference to a coordinate system. Vectors will be denoted by small letters **a**, **b**, **c**, etc. in heavy type, and may be represented by line segments $\overrightarrow{PQ}$, $\overrightarrow{QR}$, etc. (In written work the vector **a** is distinguished from the scalar (or pure number) a by a short line over the letter thus: $\bar{a}$.) Since the point of application and exact line of action of a vector is not important, either of the parallel and equal line segments $\overrightarrow{PQ}$, $\overrightarrow{SR}$, may be taken to represent the vector **a**. The *magnitude* of **a** is denoted by $|\mathbf{a}|$, and this is read as the *modulus* of **a**. *A unit vector* is one whose modulus is unity. A *null vector* is one whose modulus is zero. The vectors **a** and **b** are *equivalent* if they have equal moduli and act in the same direction. In this case we write $\mathbf{a} = \mathbf{b}$. The vector equal in magnitude and opposite in direction to **a** is written as $-\mathbf{a}$. If **a** is represented by $\overrightarrow{PQ}$, then $-\mathbf{a}$ is represented by $\overrightarrow{QP}$.

(i)

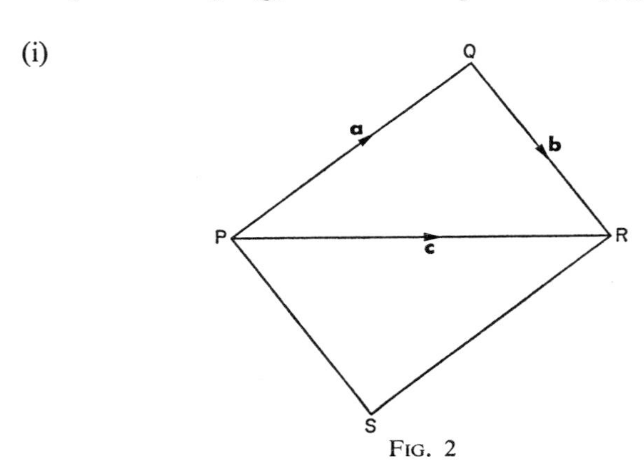

FIG. 2

In the parallelogram $PQRS$, let $\overrightarrow{PQ}$, $\overrightarrow{QR}$ represent the vectors **a**, **b** respectively.

Then $$\overrightarrow{PQ} = \overrightarrow{SR} = \mathbf{a}; \quad \overrightarrow{QR} = \overrightarrow{PS} = \mathbf{b}.$$

The displacement $\overrightarrow{PQ}$ followed by $\overrightarrow{QR}$ is equivalent to the direct displacement $\overrightarrow{PR}$.

$$\therefore \quad \overrightarrow{PQ}+\overrightarrow{QR} = \overrightarrow{PR}$$

i.e. $$\mathbf{a}+\mathbf{b} = \mathbf{c}$$

But $$\overrightarrow{PS}+\overrightarrow{SR} = \overrightarrow{PR} \text{ also.}$$

Hence $$\mathbf{b}+\mathbf{a} = \mathbf{c}$$

So, as we saw previously, $\mathbf{a}+\mathbf{b} = \mathbf{b}+\mathbf{a}$, and vector addition is *commutative*.

(ii)

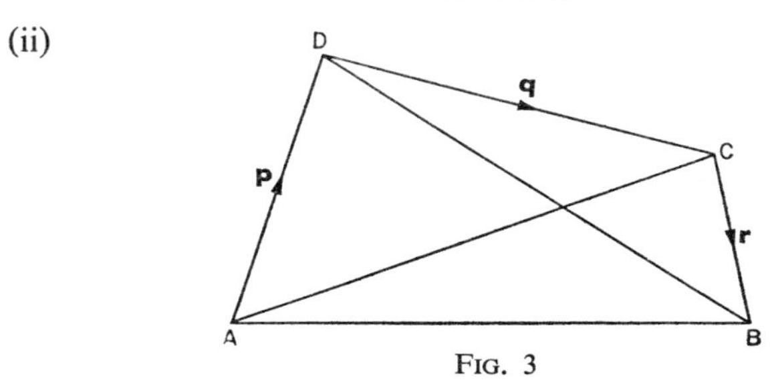

FIG. 3

Considering successive displacements $\overrightarrow{AD}$, $\overrightarrow{DB}$, we have

$$\overrightarrow{AB} = \overrightarrow{AD}+\overrightarrow{DB} = \overrightarrow{AD}+(\overrightarrow{DC}+\overrightarrow{CB})$$
$$= \mathbf{p}+(\mathbf{q}+\mathbf{r})$$

While for successive displacements $\overrightarrow{AC}$, $\overrightarrow{CB}$, we have

$$\overrightarrow{AB} = \overrightarrow{AC}+\overrightarrow{CB} = (\overrightarrow{AD}+\overrightarrow{DC})+\overrightarrow{CB}$$
$$= (\mathbf{p}+\mathbf{q})+\mathbf{r}$$

Whence $\mathbf{p}+(\mathbf{q}+\mathbf{r}) = (\mathbf{p}+\mathbf{q})+\mathbf{r}$, and vector addition is *associative.*

(iii)

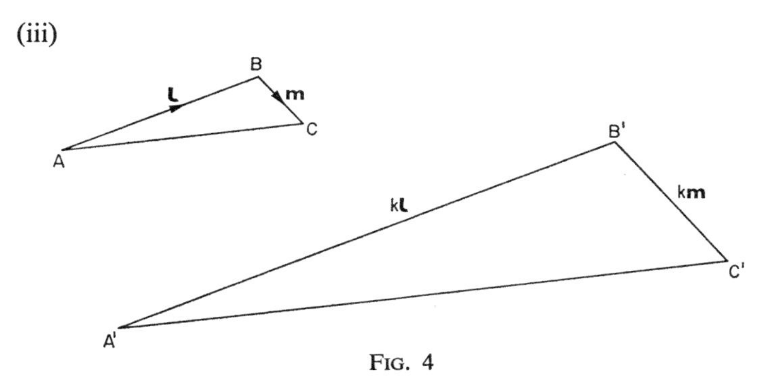

FIG. 4

If k is a scalar quantity or pure number and $\mathbf{l}$ is a vector, then $k\mathbf{l}$ is a vector parallel to the vector $\mathbf{l}$ but of modulus k times as great. Now if triangles ABC, $A'B'C'$ are similar, and $\overrightarrow{AB}$, $\overrightarrow{BC}$ are parallel to $\overrightarrow{A'B'}$, $\overrightarrow{B'C'}$ respectively, then $\overrightarrow{A'C'}$ is parallel to $\overrightarrow{AC}$ and we have:

$$\overrightarrow{AC} = \mathbf{l}+\mathbf{m}$$

Also $$\overrightarrow{A'C'} = k.\overrightarrow{AC} = k(\mathbf{l}+\mathbf{m})$$

But $$\overrightarrow{A'C'} = k\mathbf{l}+k\mathbf{m}$$

$$\therefore\ k(\mathbf{l}+\mathbf{m}) = k\mathbf{l}+k\mathbf{m},$$

so that *scalar multiplication is distributive over vector addition.* We are now in a position to apply vector methods to certain geometrical problems.

APPLICATIONS IN ELEMENTARY GEOMETRY

Examples.

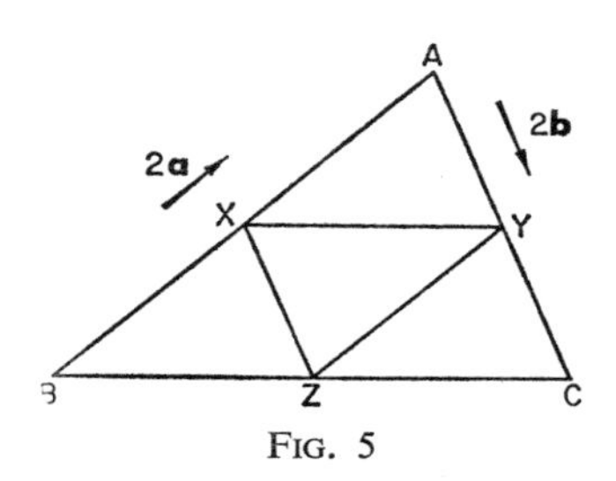

FIG. 5

(i) In the triangle ABC, X, Y, Z are midpoints of the sides AB, AC, BC, respectively. Show that the sides of triangle XYZ are respectively parallel to the sides of triangle CBA.

Let $\overrightarrow{BA}$ represent the vector $2\mathbf{a}$, and $\overrightarrow{AC}$ the vector $2\mathbf{b}$.

Then $$\overrightarrow{BX} = \overrightarrow{XA} = \mathbf{a};\ \overrightarrow{AY} = \overrightarrow{YC} = \mathbf{b}.$$

Now $$\overrightarrow{XY} = \overrightarrow{XA}+\overrightarrow{AY} = \mathbf{a}+\mathbf{b}$$

$$\overrightarrow{BC} = \overrightarrow{BA}+\overrightarrow{AC} = 2\mathbf{a}+2\mathbf{b} = 2(\mathbf{a}+\mathbf{b})$$

Hence the vector $\vec{BC}$ is parallel to the vector $\vec{XY}$ and its magnitude is twice as great, i.e. $\vec{XY}$ is parallel to $\vec{BC}$ and $2XY = BC$.

Again, $\vec{BC} = 2\mathbf{a}+2\mathbf{b}$, therefore $\vec{ZC} = \mathbf{a}+\mathbf{b}$.

$$\begin{aligned} \vec{ZY} &= \vec{ZC}+\vec{CY} \\ &= \mathbf{a}+\mathbf{b}+-\mathbf{b} \\ &= \mathbf{a} \end{aligned}$$

Hence, the vector $\vec{ZY}$ is parallel to, and half the magnitude of $\vec{BA}$, i.e. $\vec{YZ}$ is parallel to $\vec{AB}$ and $2YZ = AB$. Similarly it may be shown that XZ is parallel to AC and $2XZ = AC$.

(ii) P, R, Q are three collinear points whose positions relative to a fixed point O are given by the vectors $\mathbf{p}$, $\mathbf{r}$, $\mathbf{q}$, respectively.

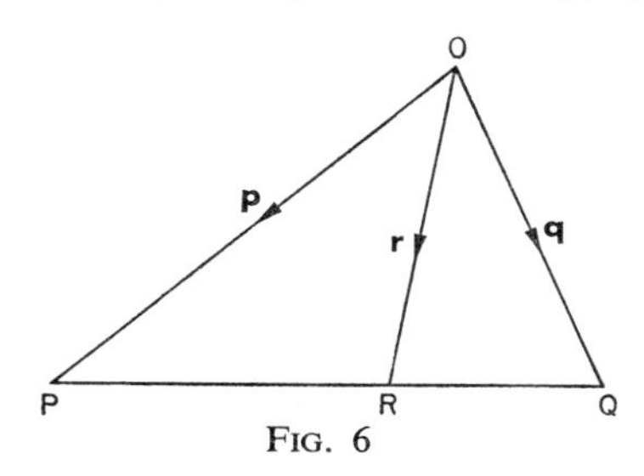

FIG. 6

If $PR:RQ = l:m$, express $\mathbf{r}$ in terms of the vectors $\mathbf{p}$ and $\mathbf{q}$.

Now $\quad \mathbf{r} = \mathbf{q}+\vec{QR}$, or $\vec{QR} = \mathbf{r}-\mathbf{q}$

and $\quad \mathbf{r} = \mathbf{p}+\vec{PR}$, or $\vec{PR} = \mathbf{r}-\mathbf{p}$

But $\quad \dfrac{PR}{RQ} = \dfrac{l}{m}$, therefore $m\vec{PR} = l\vec{RQ}$.

i.e. $$m\vec{PR}-l\vec{RQ} = 0$$

or $$m\vec{PR}+l\vec{QR} = 0$$

or $$m(\mathbf{r}-\mathbf{p})+l(\mathbf{r}-\mathbf{q}) = 0$$

i.e. $$\mathbf{r} = \frac{m\mathbf{p}+l\mathbf{q}}{m+l}$$

(iii) A, B, C are the vertices of a triangle, G is its centre of gravity (i.e. the point of intersection of its medians) and O is a fixed point. Denoting $\overrightarrow{OA}$, $\overrightarrow{OB}$, $\overrightarrow{OC}$ by vectors **a**, **b**, **c**, respectively, express $\overrightarrow{OG}$ in terms of these vectors. (**a** in this context is sometimes called the *position vector* of A relative to O.)

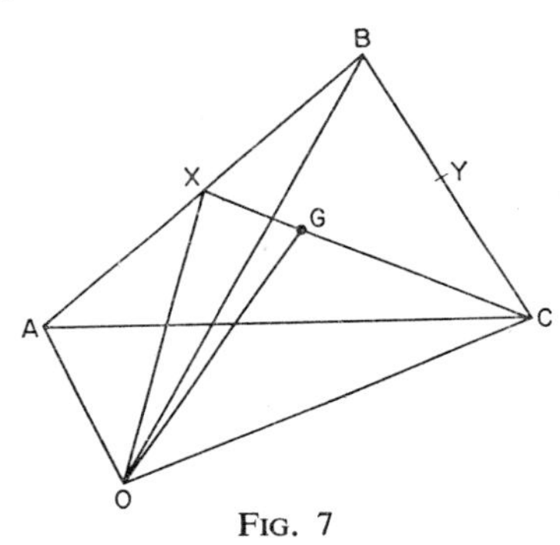

FIG. 7

If X, Y are the midpoints of the sides AB, BC, then triangles XYG, CAG are similar with $\dfrac{XG}{GC} = \frac{1}{2}$. By the result proved in (ii), the position vector of X

$$\overrightarrow{OX} = \frac{\mathbf{a}+\mathbf{b}}{2}.$$

Applying the same result again to the triangle OXC we have

$$\overrightarrow{OG} = \frac{2\overrightarrow{OX}+\overrightarrow{OC}}{2+1} = \frac{\mathbf{a}+\mathbf{b}+\mathbf{c}}{3}$$

Exercise 8(*b*)

1. Express the following vectors in terms of **p**, **q**, **r**, **s**, as denoted in Fig. 8.

(a) $\overrightarrow{AC}$ (b) $\overrightarrow{BD}$ (c) $\overrightarrow{AD}$ (d) $\overrightarrow{BE}$

(e) $\overrightarrow{AE}$

If F is the midpoint of AE find:

(f) $\overrightarrow{FB}$ (g) $\overrightarrow{FC}$ (h) $\overrightarrow{FD}$

Check your results by showing that $\overrightarrow{FB}-\overrightarrow{FD} = \overrightarrow{DB}$.

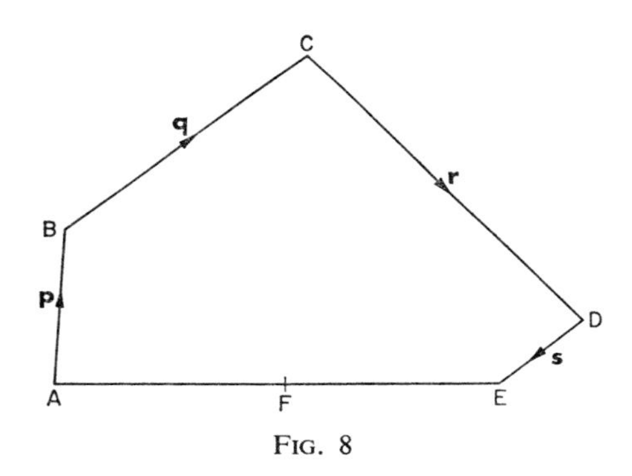

Fig. 8

2. $ABCDEF$ is a regular hexagon and O is the point of intersection of its diagonals. Express the following in terms of **a**, **b**, **c**.

(a) $\overrightarrow{DE}$ (b) $\overrightarrow{EF}$ (c) $\overrightarrow{FA}$ (d) $\overrightarrow{OC}$

(e) $\overrightarrow{OD}$ (f) $\overrightarrow{OE}$ (g) $\overrightarrow{AC}$ (h) $\overrightarrow{AD}$

(i) $\overrightarrow{AE}$

Show that $\mathbf{a}+\mathbf{c} = \mathbf{b}$.

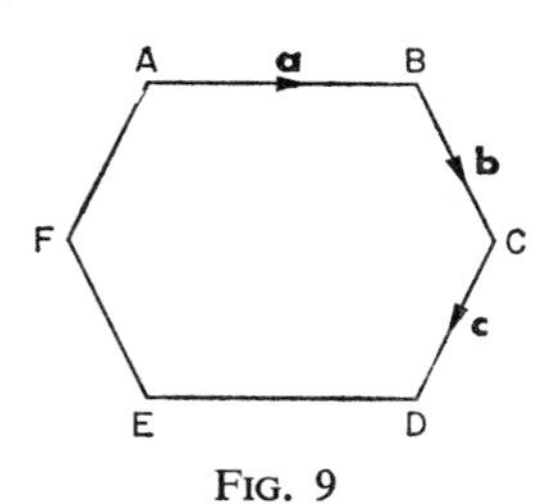

Fig. 9

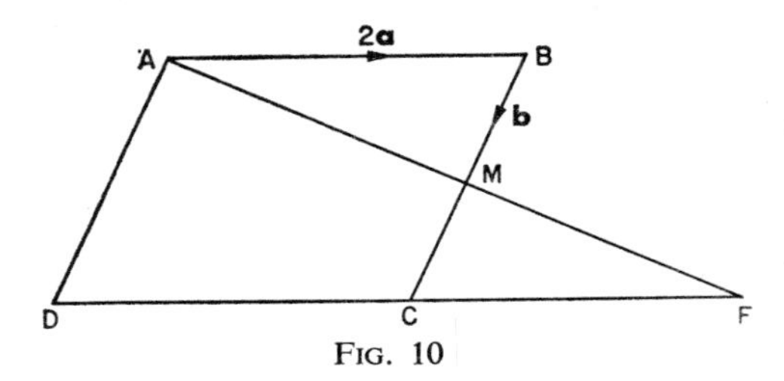

FIG. 10

3. In Fig. 10, $ABCD$ is a parallelogram and M is the midpoint of BC. If $\overrightarrow{AB} = 2\mathbf{a}$ and $\overrightarrow{BM} = \mathbf{b}$, express the following in terms of **a** and **b**.

(a) $\overrightarrow{AM}$ (b) $\overrightarrow{BC}$ (c) $\overrightarrow{AD}$ (d) $\overrightarrow{DC}$

(e) $\overrightarrow{AC}$

If AMF is a straight line and $\overrightarrow{AM} = p\mathbf{a}+q\mathbf{b}$, then $\overrightarrow{MF} = k(p\mathbf{a}+q\mathbf{b})$. State the values of p and q and show that k must be 1. Hence show that $AM = MF$. Express $\overrightarrow{AC}$ and $\overrightarrow{BF}$ in terms of **a**, **b**, and hence show that AC, BF are parallel and equal.

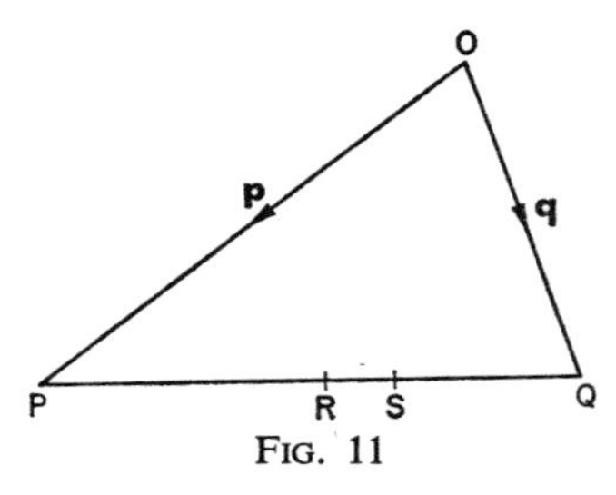

FIG. 11

4. In fig. 11, R bisects PQ and S trisects PQ. Express in terms of **p** and **q** the vectors $\overrightarrow{OR}$, $\overrightarrow{OS}$, $\overrightarrow{RS}$.

5. $ABCD$ is a square and G is the intersection of its diagonals (i.e. its centre of gravity). If the position vectors of A, B, C, D, relative to O, i.e. $\overrightarrow{OA}$, $\overrightarrow{OB}$, $\overrightarrow{OC}$, $\overrightarrow{OD}$, are denoted by **a**, **b**, **c**, **d**

respectively, express $\overrightarrow{OG}$ in terms of (i) **a** and **c**, (ii) **b** and **d**. Hence show that $\mathbf{a}+\mathbf{c} = \mathbf{b}+\mathbf{d}$. By expressing $\overrightarrow{DA}$ and $\overrightarrow{CB}$ in terms of **a**, **b**, **c**, **d**, prove the same result by a different method. Suppose now that A, B, C, D are equal weights at the vertices of a *quadrilateral*, and writing $\overrightarrow{OA}$, $\overrightarrow{OB}$, $\overrightarrow{OC}$, $\overrightarrow{OD}$ as **a**, **b**, **c**, **d** again, find the position vector of the centre of gravity of (a) A and B, (b) A, B and C, (c) A, B, C and D.

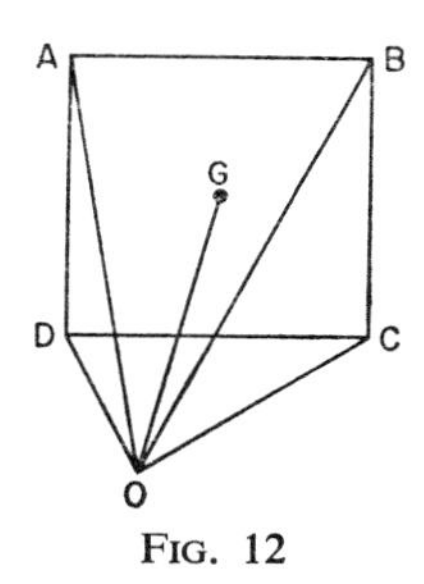

FIG. 12

6. $ABCD$ is a rectangle and $PQCD$ is a parallelogram. Prove by a vector method, that $ABQP$ is a parallelogram. (*Hint.* Denote $\overrightarrow{DA}$ by **a** and $\overrightarrow{DP}$ by **b** and then find $\overrightarrow{AP}$, $\overrightarrow{BQ}$ in terms of **a** and **b**.)

7. The sides $\overrightarrow{AB}$, $\overrightarrow{BC}$, $\overrightarrow{CA}$ of a triangle ABC are denoted by vectors **a**, **b**, **c**. Prove that $\mathbf{a}+\mathbf{b}+\mathbf{c} = 0$.

8. ABC is a triangle and D, E, F are the midpoints of its sides. If the medians intersect at O, prove that $\overrightarrow{OD}+\overrightarrow{OE}+\overrightarrow{OF} = \overrightarrow{OA}+\overrightarrow{OB}+\overrightarrow{OC}$.

9. $ABCD$ is a quadrilateral. M is the midpoint of BD and N the midpoint of AC. Show that $\overrightarrow{AB}+\overrightarrow{AD}+\overrightarrow{CB}+\overrightarrow{CD} = 4\overrightarrow{NM}$.

10. $ABCD$ is a quadrilateral. By taking the position vectors of A, B, C and D relative to a fixed point O as **a**, **b**, **c**, **d** respectively, obtain the position vectors of the midpoints of the sides, and hence show that these midpoints form the vertices of a parallelogram.

11. If the diagonals of a quadrilateral $ABCD$ are equal in length, show that the midpoints of the sides form the vertices of a rhombus.

12. In the parallelogram $ABCD$, P and Q lie on the diagonal BD so that $BQ = PD$. Show, by a vector method, that $AQCP$ is a parallelogram.

13. Use the result of Example (ii) to show that if the position vectors $\mathbf{p}$, $\mathbf{q}$, $\mathbf{r}$ of three points P, Q, R satisfy the equation $a\mathbf{p}+b\mathbf{q}+c\mathbf{r} = 0$, where $a+b+c = 0$, then the points P, Q, R are collinear.

14. Show that the vectors of the form $l\mathbf{a}+m\mathbf{b}$, where l and m are real numbers, form an infinite group under addition. State the identity element and the inverse of $l\mathbf{a}+m\mathbf{b}$.

MORE APPLICATIONS OF THE VECTOR LAW OF ADDITION—VECTOR QUANTITIES

By a vector quantity we mean one in which both the magnitude and the direction are important. (Strictly speaking they are quantities which, when represented by a vector, obey the vector law of addition; but as this is the case in all the applications dealt with we shall not labour the point.)

As we have seen, *displacement* is a vector quantity. In order to get from London to Manchester it is not sufficient to say that one must displace oneself 184 miles; the displacement must take place in a particular direction. *Velocity* and *acceleration* are also vector quantities. No car will win the Monte Carlo Rally simply by maintaining a high speed; the direction in which the car moves is an equally vital factor. *Momentum and impulse* are vector quantities. A footballer may have a strong kick, but in order to score he must take care in which direction he applies an impulse to the ball. *Force* is a vector quantity, but rather a complicated one. It is unlike the other vector quantities mentioned in that not only must its magnitude and direction be specified, but also *the line in which its action takes place.* We may indeed apply the triangle law of addition to forces which act at a point in order to determine their resultant. Our answer, however, will only give us information about its magnitude and direction; it will not give us the line along which the force acts. Of course, if two forces act through a point, their resultant acts through the same point, but by stating this we are actually invoking a new principle—the law of moments. At a more advanced stage it is possible to show that a force can only be completely specified if we use *two* different free vectors. Kinetic energy, work, quantity of heat, etc., are *not* vector quantities.

Example (i). An aircraft travels at 200 knots in still air. If a wind of 40 knots blows steadily from the south west and the pilot wishes to fly due north, find the course he must set and the ground speed of the aircraft. On the return journey the wind speed and direction are unaltered. Find the new course and ground speed.

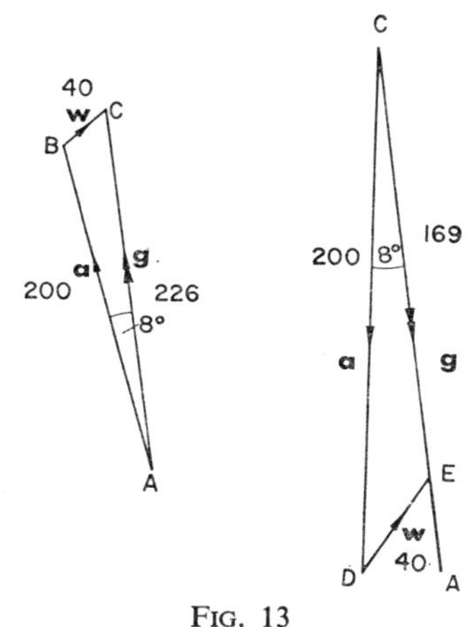

FIG. 13

The ground speed (represented by vector **g**) is the speed at which the aircraft travels relative to a stationary observer on the ground. It is the resultant of the air speed (**a**) and the wind speed (**w**), i.e. $\mathbf{a}+\mathbf{w} = \mathbf{g}$. Thus, although the plane actually travels along the line AC, we can think of it travelling in still air from A to B, and then being blown by the wind from B to C.

In practice this process is continuous; the plane travels from A to C but its "nose" points in the direction $\overrightarrow{AB}$ throughout the outward journey. The triangles of velocity for outward and inward journeys are shown in Fig. 13, together with the courses and ground speeds as obtained by geometrical construction.

Example (ii). A ball hits a rectangular corner and receives a blow or impulse from each wall as shown in Fig. 14. In which direction does it rebound?

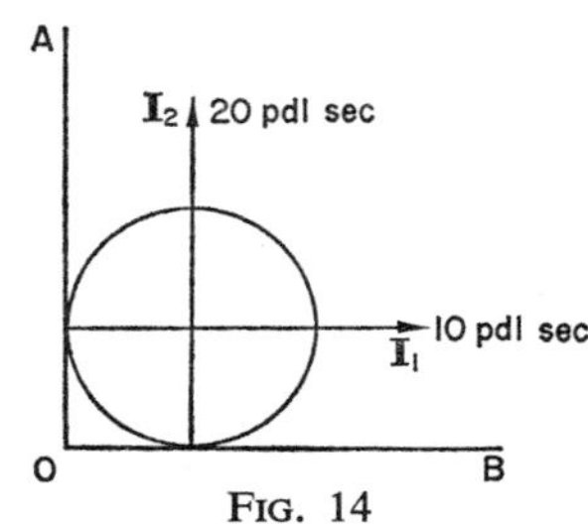

FIG. 14

The resultant impulse (vector **I**) is obtained from the two perpendicular impulses (represented by vectors $\mathbf{I}_1$, $\mathbf{I}_2$) by the vector law of addition. Thus $\mathbf{I} = \mathbf{I}_1 + \mathbf{I}_2$. The magnitude and direction of **I** may be obtained by scale drwing or by the use of simple trigonometry and the use of Pythagoras' theorem. The angle of rebound with $\overrightarrow{OA}$ is the angle whose tangent is $\frac{10}{20}$, i.e. 26° 34′. The magnitude of the impulse is $\sqrt{(20^2+10^2)}$, i.e. 22·36 pdl secs.

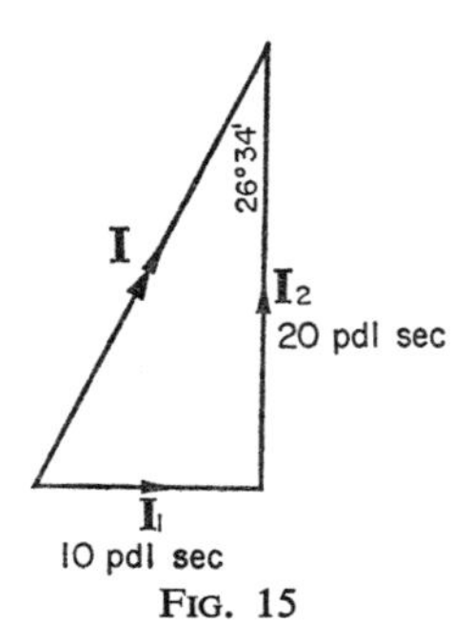

FIG. 15

Example (iii). Three men are pushing a broken down car, and by exerting forces as shown in Fig. 16 they just keep it moving. A fourth man comes along and boasts that he can do the job single-handed. Wisely, the others allow him to do so. In what direction and with what force must he push?

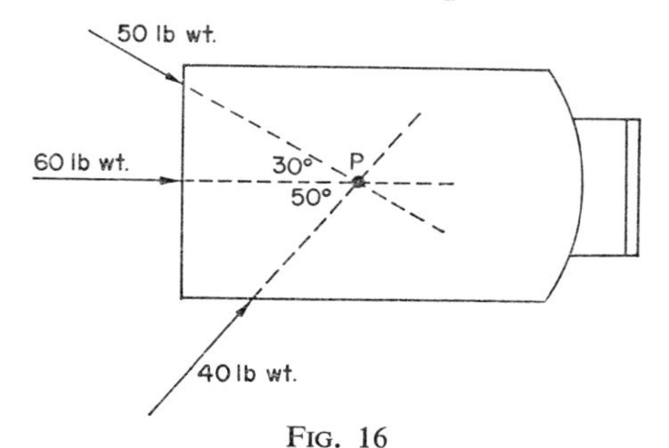

FIG. 16

If we represent the forces exerted by the three men by the vectors **a**, **b**, **c** respectively, then the force to be exerted single-handedly by the fourth man (vector **d**) is given by $\mathbf{d} = \mathbf{a}+\mathbf{b}+\mathbf{c}$. From the scale drawing of the force polygon (Fig. 17) we see that $|\mathbf{d}| = 129$ lb wt., and the direction of **d** makes an angle of approximately $2\frac{1}{2}°$ with the line of travel of the vehicle. Notice that Fig. 17 does

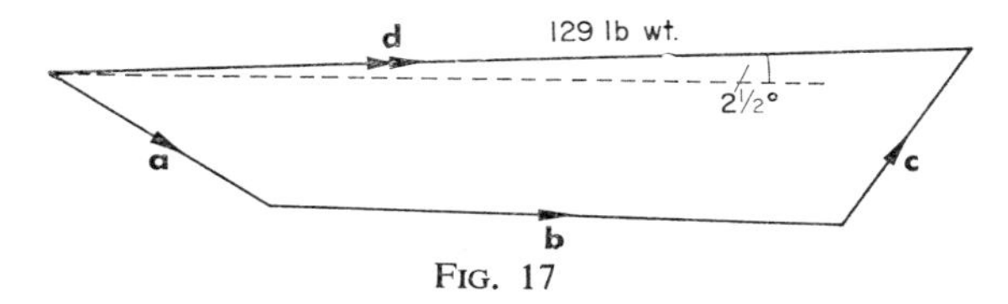

FIG. 17

not tell us at what point on the car the fourth man must apply this resultant force. If, however, the lines of action of the 50-lb, 60-lb and 40-lb forces all pass through the point P (Fig. 16), then, by the law of moments, the line of action of their resultant must also pass through this point. (It might be thought that the resultant should act in the direction of travel of the vehicle. The result obtained here is, however, quite feasible if other factors, such as the camber of the road, are operating.)

Exercise 8(*c*)

1. A man walks at 4 m.p.h. relative to the deck from the port to the starboard bow of a ship which is steaming due west at 12 m.p.h. With what speed and in which direction is the man actually travelling?

2. In Question 1, a north wind of 6 m.p.h. is blowing. Find the direction of the smoke trail from the ship's funnel.

3. A man can swim at 3 m.p.h. in still water. Points A and B on opposite sides of a river, which is $\frac{1}{4}$ mile wide and flowing at 1 m.p.h., are such that AB is perpendicular to the parallel banks.

(a) If the man tries to swim in the direction $\overrightarrow{AB}$, how far from B will he land and what time will he take? What will be his actual speed?

(b) In what direction should he attempt to swim from A (i.e. what "course" should he "set"), in order to land at B? What would his time and actual speed be in this case?

4. OBA is a spoke of a wheel which is rotating with constant angular acceleration. OA is 1 ft, B is the midpoint of OA, and the point A has a linear acceleration of $\frac{1}{3}$ ft per second per second and a velocity of $\frac{1}{2}$ ft per second. A beetle hurries along $\overrightarrow{AO}$ and when it reaches B its acceleration and velocity *along the spoke* are 2 inches per second per second and 4 inches per second respectively. Find the actual velocity and acceleration of the beetle.

5. A man walks along country lanes from A to D. His route is as follows: 4 miles N 60° E, 3 miles due north, and finally 2 miles north-east. A crow flies directly from A to D. Find the direction and length of its flight. If the man walks at 4 m.p.h. and the crow flies at 12 m.p.h., how much less time does the crow take?

6. An electron beam is directed along an electric field of force 20 units. If another field of force of 5 units now acts across the beam and in a direction perpendicular to the first, find the angle of deflection of the beam. If a third field of force of magnitude 10 units now acts across the beam and in a direction which is perpendicular to each of the other two fields of force, find the new deflection of the beam from its original direction.

7. The places A, B, C, D form a square of side 100 miles. The direction of $\overrightarrow{AB}$ is due east, and a wind of 50 knots blows steadily from the north west. An aircraft, which travels at 200 knots in still air, flies from A along the four sides of the square in turn. Find the total time for the round trip. By how much does this exceed or fall short of the time taken on a perfectly calm day?

THE SCALAR PRODUCT OF TWO VECTORS

The scalar product of two vectors **a** and **b** is defined as the product of their moduli multiplied by the cosine of the angle θ between their directions. We write the product as $\mathbf{a}.\mathbf{b}$.

Thus $$\mathbf{a}.\mathbf{b} = ab\cos\theta = b.a\cos\theta = \mathbf{b}.\mathbf{a}$$

For the scalar product $\mathbf{a}.\mathbf{a}$, $\theta = 0°$ and $\cos 0° = 1$.

Hence $$\mathbf{a}.\mathbf{a} = a^2.$$

If **a** and **b** are perpendicular, then in this case

$$\theta = \frac{\pi}{2} \text{ (or } 90°\text{) and } \cos 90° = 0.$$

Hence $$\mathbf{a}.\mathbf{b} = 0.$$

Conversely, if the scalar product of two non zero vectors is zero, the vectors are perpendicular.

Scalar multiplication is obviously a commutative operation. We can show that it also obeys the distributive law.

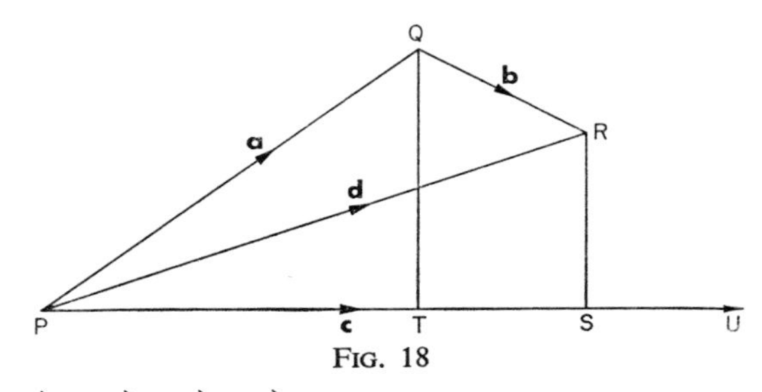

FIG. 18

Denote $\vec{PQ}$, $\vec{QR}$, $\vec{PU}$, $\vec{PR}$ by **a**, **b**, **c**, **d** respectively. Then

$$(\mathbf{a}+\mathbf{b}).\mathbf{c} = \mathbf{d}.\mathbf{c} = PR.PU \cos RPU = PS.PU.$$

Also $\quad \mathbf{a}.\mathbf{c} = PT.PU$

and $\quad \mathbf{b}.\mathbf{c} = TS.PU.$

Hence $\quad \mathbf{a}.\mathbf{c}+\mathbf{b}.\mathbf{c} = (PT+TS)PU = PS.PU.$

Therefore $(\mathbf{a}+\mathbf{b}).\mathbf{c} = \mathbf{a}.\mathbf{c}+\mathbf{b}.\mathbf{c}.$

We shall see in what follows that by defining the scalar product of two vectors in this way certain geometrical proofs become very simple indeed. Further, the definition is exactly equivalent to the rule of multiplication for a row matrix or vector and a column matrix or vector.

FURTHER APPLICATIONS TO ELEMENTARY GEOMETRY

Example (i). The Cosine Rule and Pythagoras' Theorem.

In Fig. 19 we have: $\quad \mathbf{a}+\mathbf{b} = \mathbf{c}$

Squaring both sides (i.e. applying the distributive law of multiplication),

$$(\mathbf{a}+\mathbf{b}).(\mathbf{a}+\mathbf{b}) = \mathbf{c}.\mathbf{c}$$

i.e. $\quad \mathbf{c}.\mathbf{c} = \mathbf{a}.\mathbf{a}+\mathbf{b}.\mathbf{b}+2\mathbf{a}.\mathbf{b}$

$$\therefore\ c^2 = a^2+b^2+2ab\cos(180^\circ - C)$$

or $\quad c^2 = a^2+b^2-2ab\cos C$

If $\angle C = 90°$, $\cos C = 0$ and we have Pythagoras' theorem $c^2 = a^2+b^2$.

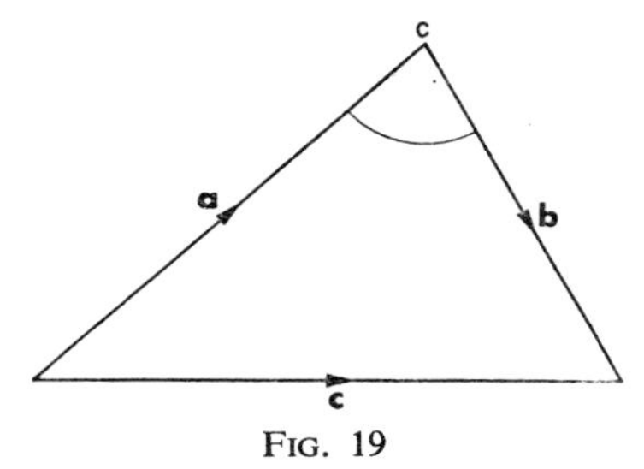

FIG. 19

Example (ii). If a line is drawn from the centre of a circle to the midpoint of a chord, the line is perpendicular to the chord.
Proof.

$$\vec{OM}+\vec{MA} = \vec{OA}$$

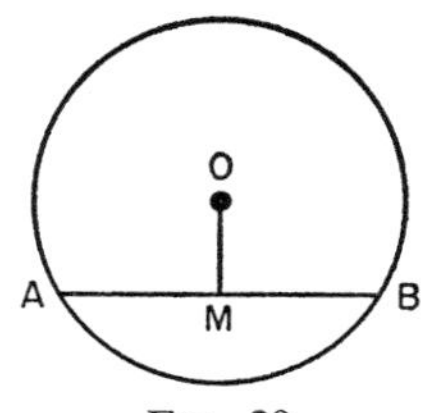

FIG. 20

Squaring scalarly

$$OM^2+2\vec{OM}.\vec{MA}+MA^2 = OA^2 \qquad 1$$

$$\vec{OM}+\vec{MB} = \vec{OB}$$

or

$$\vec{OM}-\vec{BM} = \vec{OB}$$

$$\therefore\ OM^2-2\vec{OM}.\vec{BM}+BM^2 = OB^2 \qquad 2$$

Subtracting 2 from 1 and remembering that $\vec{MA} = \vec{BM}$, we have

$$4\vec{OM}.\vec{MA} = 0$$

$$\therefore\ \vec{OM} \text{ is perpendicular to } \vec{MA}.$$

Exercise 8(*d*)

1. Prove by a vector method that the angle in a semi-circle is a right angle.

(Take AOB as diameter and C as any point on the circumference. $\overrightarrow{CA} = \overrightarrow{CO} + \overrightarrow{OA}$. Find a similar expression for $\overrightarrow{CB}$ and hence show that the scalar product of $\overrightarrow{CA}$ and $\overrightarrow{CB}$ is zero.)

2. $ABDE$ is a trapezium with AB parallel to ED. C is a point on BD such that $AB = BC$, $CD = DE$.
Let $\mathbf{i}$, $\mathbf{n}$ be unit vectors in the directions $\overrightarrow{AB}$, $\overrightarrow{BC}$ respectively, and let $\overrightarrow{BC}$, $\overrightarrow{CD}$ have moduli k, n respectively. Show that $\overrightarrow{AC} = k(\mathbf{i}+\mathbf{n})$ and find $\overrightarrow{EC}$. By forming the scalar product $\overrightarrow{AC}.\overrightarrow{EC}$ prove that $\angle ACE = 90°$.

3. Triangle ABC is right angled at B and X is the midpoint of AC. Taking the position vectors of A and C relative to B as $\mathbf{a}$ and $\mathbf{c}$, express $\overrightarrow{BX}$ and $\overrightarrow{AX}$ in terms of $\mathbf{a}$ and $\mathbf{c}$ and hence show that $AX = BX$.

4. Two circles touch externally at A. A common tangent touches the circles at B and C respectively. Prove by a vector method that $\angle BAC$ is a right angle.

5. $ABCD$ is a tetrahedron with AB perpendicular to CD, AD perpendicular to BC. Prove that AC is also perpendicular to BD. (*Hint.* Denote $\overrightarrow{AB}$, $\overrightarrow{AC}$, $\overrightarrow{AD}$ by $\mathbf{a},\mathbf{c},\mathbf{b}$ respectively. Find $\overrightarrow{CD}$, $\overrightarrow{BC}$ in terms of $\mathbf{a}$, $\mathbf{c}$, $\mathbf{b}$, and express the data as $\mathbf{a}.\overrightarrow{CD} = 0$, $\mathbf{b}.\overrightarrow{BC} = 0$. Add these equations.)

6. ABC is any triangle. Squares AF, AJ, BD are drawn on AB, AC and BC respectively. Prove by a vector method that $\overrightarrow{FC}$ is perpendicular to $\overrightarrow{AE}$. (E is the fourth corner of the square BD.)

7. A particle is moved from a point O to a point P by the action of a force $\mathbf{F}$ whose line of action is inclined at an angle θ to $\overrightarrow{OP}$. Show that the work done by the force is $\mathbf{F}.\mathbf{d}$, where $\mathbf{d} = \overrightarrow{OP}$.

UNIT VECTORS AND CO-ORDINATE GEOMETRY

Many simple results in coordinate geometry may be obtained by vector methods.

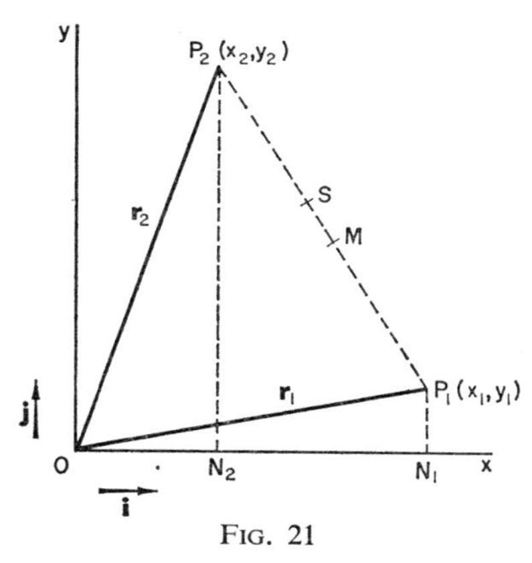

FIG. 21

Let $\mathbf{i}$, $\mathbf{j}$ be unit vectors along $\overrightarrow{OX}$, $\overrightarrow{OY}$ respectively.

Then in Fig. 21 we have:

$$\overrightarrow{ON_1} = \mathbf{i}.ON_1 = \mathbf{i}x_1$$

$$\overrightarrow{ON_2} = \mathbf{i}.ON_2 = \mathbf{i}x_2$$

$$\overrightarrow{N_1P_1} = \mathbf{j}.N_1P_1 = \mathbf{j}y_1$$

$$\overrightarrow{N_2P_2} = \mathbf{j}.N_2P_2 = \mathbf{j}y_2$$

Also,

$$\overrightarrow{OP_1} = \overrightarrow{ON_1} + \overrightarrow{N_1P_1}$$

$$\overrightarrow{OP_2} = \overrightarrow{ON_2} + \overrightarrow{N_2P_2}$$

or

$$\mathbf{r}_1 = \mathbf{i}x_1 + \mathbf{j}y_1$$

$$\mathbf{r}_2 = \mathbf{i}x_2 + \mathbf{j}y_2$$

These are the *position vectors* of P_1 and P_2 relative to O.

$$\text{Now} \qquad \overrightarrow{P_1P_2} = \mathbf{r}_2 - \mathbf{r}_1$$
$$= \mathbf{i}(x_2-x_1)+\mathbf{j}(y_2-y_1)$$

$$\text{Hence} \qquad P_1P_2 = \sqrt{[(x_2-x_1)^2+(y_2-y_1)^2]}$$

and the gradient of $P_1P_2 = (y_2-y_1)/(x_2-x_1)$.

If M is the midpoint of P_1P_2, then by a previous result

$$\overrightarrow{OM} = \tfrac{1}{2}(\overrightarrow{OP_1}+\overrightarrow{OP_2})$$
$$= \mathbf{i}\,\frac{(x_1+x_2)}{2}+\mathbf{j}\,\frac{(y_1+y_2)}{2}$$

i.e. the coordinates of M are $\left(\dfrac{x_1+x_2}{2}, \dfrac{y_1+y_2}{2}\right)$.

If S divides P_1P_2 so that $P_1S:SP_2 = l:m$, then by the same result

$$(l+m)\overrightarrow{OS} = l\mathbf{r}_2+m\mathbf{r}_1$$

i.e.

$$\overrightarrow{OS} = \frac{(lx_2+mx_1)\mathbf{i}+(ly_2+my_1)\mathbf{j}}{(l+m)}$$

i.e. the coordinates of S are $\left(\dfrac{lx_2+mx_1}{l+m}, \dfrac{ly_2+my_1}{l+m}\right)$. If, however, S is *any* point on the line P_1P_2 with position vector, say, $\mathbf{r}$, then,

$$a\mathbf{r}_1+b\mathbf{r}_2+c\mathbf{r} = 0$$

where

$$a+b+c = 0.$$

More simply we may write:

$$\mathbf{r} = t_1\mathbf{r}_1+t_2\mathbf{r}_2 \text{ where } t_1+t_2 = 1.$$

This is the vector equation of the line joining points with position vectors $\mathbf{r}_1, \mathbf{r}_2$.

If $r = \mathbf{i}x+\mathbf{j}y$ we can put this result into a more familiar form, for

$$\mathbf{i}x+\mathbf{j}y = t_1(\mathbf{i}x_1+\mathbf{j}y_1)+t_2(\mathbf{i}x_2+\mathbf{j}y_2).$$

So comparing the coefficients of **i** and **j**, and using $t_1 = 1-t_2$,

$$x = x_1+t_2(x_2-x_1)$$
$$y = y_1+t_2(y_2-y_1).$$

Or, eliminating t, $$\frac{y-y_1}{y_2-y_1} = \frac{x-x_1}{x_2-x_1},$$

which is the equation of the line joining the points (x_1,y_1) (x_2,y_2).

We are now in a position to see the significance of our definition of the scalar product of two vectors. For

$$\mathbf{r}_1 . \mathbf{r}_2 = (\mathbf{i}x_1+\mathbf{j}y_1).(\mathbf{i}x_2+\mathbf{j}y_2)$$
$$= x_1x_2+y_1y_2 \text{ since } \mathbf{i}.\mathbf{i} = 1 \text{ and } \mathbf{i}.\mathbf{j} = 0.$$

But this is the matrix product of the row vector (x_1,y_1) and the column vector $\begin{pmatrix} x_2 \\ y_2 \end{pmatrix}$, i.e.

$$(x_1 y_1)\begin{pmatrix} x_2 \\ y_2 \end{pmatrix} = x_1x_2+y_1y_2.$$

Example. A, B, C, D are the points (1,2) (3,4) (5,5) (6,0) respectively. Show that AC is perpendicular to BD. Find the angle BAC and the point P dividing AC in the ratio 3:1.

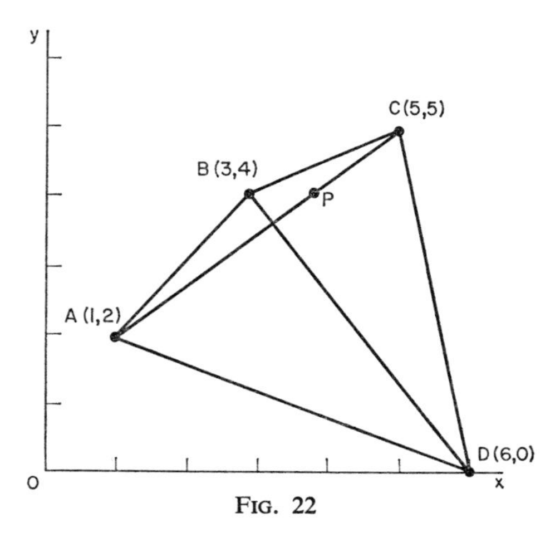

FIG. 22

First method

$$\vec{OA} = \mathbf{i}+2\mathbf{j}$$

$$\vec{OC} = 5\mathbf{i}+5\mathbf{j}$$

$$\therefore \quad \vec{AC} = \vec{OC}-\vec{OA}$$

$$= 4\mathbf{i}+3\mathbf{j}$$

$$\vec{OD} = 6\mathbf{i}$$

$$\vec{OB} = 3\mathbf{i}+4\mathbf{j}$$

$$\therefore \quad \vec{BD} = \vec{OD}-\vec{OB}$$

$$= 3\mathbf{i}-4\mathbf{j}$$

The scalar product $\vec{AC}.\vec{BD} = (4\mathbf{i}+3\mathbf{j}).(3\mathbf{i}-4\mathbf{j})$

$$= 12-12$$

$$= 0$$

Hence AC is perpendicular to BD.

$$\vec{AB} = \vec{OB}-\vec{OA} = 2\mathbf{i}+2\mathbf{j}$$

$$\vec{AC} = \vec{OC}-\vec{OA} = 4\mathbf{i}+3\mathbf{j}$$

$$\therefore \quad AB = \sqrt{(2^2+2^2)} = 2\sqrt{2}$$

and $$AC = \sqrt{(3^2+4^2)} = 5$$

Now $$\vec{AB}.\vec{AC} = AB.AC\cos\angle BAC$$

i.e. $$(2\mathbf{i}+2\mathbf{j}).(4\mathbf{i}+3\mathbf{j}) = 2\sqrt{2}.5.\cos BAC$$

$$\therefore \quad \cos BAC = \frac{14}{10\sqrt{2}} = \frac{7\sqrt{2}}{10} \text{ or } \cdot 9899 \text{ (4 d.p.)}$$

$$\therefore \quad \angle BAC = 8°8'.$$

To find $\overrightarrow{OP}$ we have:

$$(3+1)\overrightarrow{OP} = 1.\overrightarrow{OA}+3.\overrightarrow{OC}$$

$$\therefore\ 4\overrightarrow{OP} = (\mathbf{i}+2\mathbf{j})+3(5\mathbf{i}+5\mathbf{j})$$

$$= 16\mathbf{i}+17\mathbf{j}$$

$$\therefore\ \overrightarrow{OP} = 4\mathbf{i}+4\tfrac{1}{4}\mathbf{j}$$

$\therefore$ P is the point $(4, 4\frac{1}{4})$.

Second method

It is permissible to use matrix notation throughout.

Thus $$\overrightarrow{AB} = (3,4)-(1,2) = (2,2)$$

$$\overrightarrow{AC} = (5,5)-(1,2) = (4,3)$$

$$\overrightarrow{BD} = (6,0)-(3,4) = (3,-4)$$

$$\therefore\ \overrightarrow{AC}.\overrightarrow{BD} = (4,3)\begin{pmatrix}3\\-4\end{pmatrix} = 0$$

$\therefore$ AC is perpendicular to BD.

$$\overrightarrow{AB}.\overrightarrow{AC} = (2,2)\begin{pmatrix}4\\3\end{pmatrix} = 14 = 2\sqrt{2}.5.\cos BAC$$

$$\therefore\ \cos BAC = 7\sqrt{2}/10.$$

Finally, $$(3+1)\overrightarrow{OP} = 1(1,2)+3(5,5) = (16,17)$$

$$\therefore\ \overrightarrow{OP} = (4, 4\tfrac{1}{4})$$

i.e. P is the point $(4, 4\frac{1}{4})$.

Exercise 8(*e*)

Questions 1–7 refer to Fig. 1 in which unit vectors **i**, **j** act along $\overrightarrow{OX}$, $\overrightarrow{OY}$ respectively.

1. Express the following vectors in the form $a\mathbf{i}+b\mathbf{j}$:

(a) $\overrightarrow{OA}$ (b) $\overrightarrow{OC}$ (c) $\overrightarrow{OB}$ (d) $\overrightarrow{AC}$
(e) $\overrightarrow{OH}$ (f) $\overrightarrow{BD}$ (g) $\overrightarrow{OE}$ (h) $\overrightarrow{EC}$
(i) $\overrightarrow{CD}$ (j) $\overrightarrow{DG}$

2. Find by a vector method the coordinates of the point which
(a) bisects *CB*,
(b) divides *AB* in the ratio 2:1,
(c) divides *CI* in the ratio 5:1,
(d) divides *BG* in the ratio 4:3,
(e) divides *FI* in the ratio 5:7.

3. Give the vector equation of the line joining the points:
(a) *A* and *B* (b) *F* and *B* (c) *G* and *I*
(d) *H* and *I* (e) *C* and *D*

4. Evaluate the scalar products:
(a) $\overrightarrow{OA}.\overrightarrow{OC}$ (b) $\overrightarrow{IH}.\overrightarrow{IJ}$ (c) $\overrightarrow{OB}.\overrightarrow{GI}$

5. Show that the following pairs of vectors are perpendicular:
(a) $\overrightarrow{AC}$, $\overrightarrow{OB}$ (b) $\overrightarrow{OH}$, $\overrightarrow{BD}$ (c) $\overrightarrow{OE}$, $\overrightarrow{EC}$
(d) $\overrightarrow{CD}$, $\overrightarrow{DG}$

6. By using the scalar product, determine the following:
(a) $\angle AOC$ (b) $\angle BOH$ (c) $\angle ACD$
(d) $\angle ICJ$

7. Find the centre of gravity, or point of intersection of the medians, of the following:
(a) triangle *ABC* (b) triangle *FBI* (c) triangle *GCJ*

8. Unit vectors **i**, **j**, **k** act along the three mutually perpendicular axes $\overrightarrow{OX}$, $\overrightarrow{OY}$, $\overrightarrow{OZ}$. If P is the point (x, y, z) we write $\overrightarrow{OP} = \mathbf{i}x+\mathbf{j}y+\mathbf{k}z$. Evaluate:

(a) $\mathbf{i}^2$ (b) $\mathbf{j}^2$ (c) $\mathbf{k}^2$
(d) $\mathbf{i}.\mathbf{j}$ (e) $\mathbf{i}.\mathbf{k}$ (f) $\mathbf{j}.\mathbf{k}$

A, B, C, D are the points $(4,6,3)$, $(3,1,1)$, $(5,2,0)$, $(2,1,4)$ respectively. Show by a vector method that $\overrightarrow{AB}$ and $\overrightarrow{CD}$ are perpendicular. If E is the point $(5,11,5)$, show that A, B, E are collinear and that A is the midpoint of BE.

9. A, B, C are the points $(0,2,5)$, $(2,0,7)$, $(1,2,0)$ respectively and O is the origin. Find the resultant of the forces represented in magnitude and direction by the vectors $\overrightarrow{OA}$, $\overrightarrow{OB}$, $\overrightarrow{OC}$.

10. Draw a unit cube having one vertex at the origin and three coterminous edges along the axes OX, OY, OZ. Use a vector method to determine the acute angle between any two of its diagonals.

9
PROBABILITY AND STATISTICS

WE HAVE only to open a newspaper or watch the television to see facts and figures relating to such questions as unemployment in the North, the state of the political parties, the percentage of people who use a certain type of soap powder, or the relative appeal of the latest "pop" discs.

In this chapter we shall consider the ways in which information of this kind is collected, represented and analysed.

COLLECTION OF DATA

Figures relating to matters of a serious nature, such as the numbers of people in certain age groups, export figures for the motor car industry, unemployment in the coal industry, etc., are collected regularly by the Central Statistical Office and published by the Stationery Office. The industries and firms concerned make regular returns on official questionnaire forms and from the information acquired, and the patterns and trends revealed by its analysis, the Government is able to determine various aspects of its policy.

Information for Gallup polls is usually collected orally by teams of investigators who question a small but representative section of the public. Many commercial enterprises keep a close eye on the popularity and success of their products by employing door to door canvassers, or by asking a representative selection of the public to complete questionnaires. On a more technical level, many firms maintain quality control over their products by the regular testing of small sample batches. In this way unobtrusive small manufacturing faults are revealed and corrected before large scale loss and wastage occur.

Not all claims based on statistical evidence are reliable. The mathematician is genuinely concerned to arrive at the truth, but it is not unknown for extravagant commercial claims to be based on insufficient or biased figures.

The data itself may sometimes be false. In fact the design of a questionnaire is a job requiring considerable skill. Questions must be completely unambiguous and worded in such a way that a truthful answer will not compromise or incriminate the person who is answering. Even when the data is accurate and reliable it is, of course, possible to represent it in a misleading fashion. The deliberate exaggeration of scales or a curious choice of origin on the axes of graphs, or distortion in the relative size of pictorial units, can over emphasize or disguise a trend or preference revealed in the figures.

In this chapter we are mainly concerned with correct and reliable ways of looking at statistical information, but it is well for any future citizen to be aware of the possible abuses of the subject.

REPRESENTATION OF DATA—PIE CHARTS AND BAR CHARTS

Information given "raw" as a table of figures or readings, although the most straightforward method, tends to lack impact. Presented pictorially or diagrammatically it is often possible to see the whole situation at a glance.

Consider the following household budget of a man earning £10 per week.

	£	*s.*	*d.*
Food	4	0	0
Rent	2	0	0
Clothing	1	0	0
Fuel		15	0
Other expenses	1	5	0
Savings	1	0	0

These figures are in the ratio 16:8:4:3:5:4, and hence the quantities may be represented by circular sectors of angle 144°, 72°, 36°, 27°, 45°, 36°, as shown in Fig. 1. Such a diagram is called a *pie chart*.

FIG. 1

Alternatively they may be represented by columns of equal width and of heights which are in the ratio 16:8:4:3:5:4. This sort of diagram, which is commonly used to show the variations of rainfall, trade, etc., over the months of the year, is called a *bar chart*.

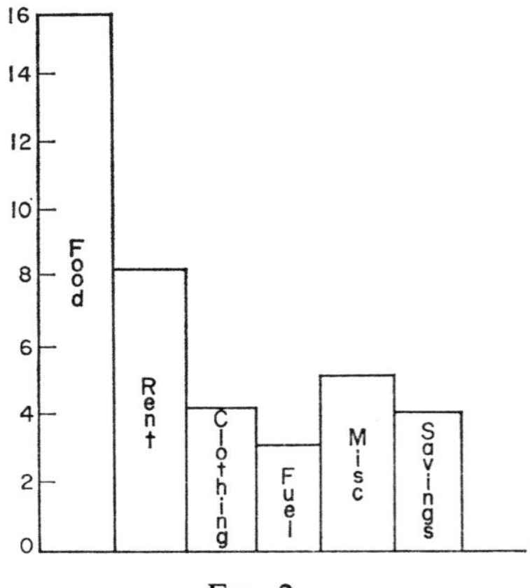

FIG. 2

Exercise 9(*a*)

Represent each of the following sets of quantities by (a) a pie chart, (b) a bar chart, (c) a purely pictorial lay-out of your own devising.

1. The average weekly fuel and lighting expenses of a certain household are:

	s.	*d.*
Coal	10	0
Coke	7	6
Electricity	12	6

2. A T.V. station broadcasts $7\frac{1}{2}$ hours each day and, on average, a programme analysis shows:

Drama	$2\frac{1}{2}$ hours
Documentary	1 hour
Sport	$1\frac{1}{2}$ hours
News	$\frac{1}{2}$ hour
Variety	2 hours.

3. The total votes polled (to the nearest 100) in a certain by-election were distributed among the various candidates as follows:

Conservative	13,500
Labour	10,500
Liberal	4,500
Independent	1,500

4. The area of land (in units of 1,000 acres) in Great Britain is shared as follows:

England	32,212
Wales	5,130
Scotland	19,463

5. The annual local authority grant to a certain school is itemized as shown.

	£	s.	d.
Building and repairs	400	0	0
Apparatus and furniture	600	0	0
Books	500	0	0
Stationery	300	0	0

FREQUENCY POLYGONS — HISTOGRAMS

More interesting are those enquiries in which the distribution of one quantity depends upon another. Consider the following quarterly sales figures relating to men's shoes in a certain department store.

Shoe size	7	$7\frac{1}{2}$	8	$8\frac{1}{2}$	9	$9\frac{1}{2}$	10	$10\frac{1}{2}$	11	$11\frac{1}{2}$
No. of pairs sold	20	24	33	37	48	61	50	35	25	17

The situation may be seen at a glance if we represent the data graphically as shown in Fig. 3. In this diagram the ordinates represent the number of pairs of each size sold, or the frequency

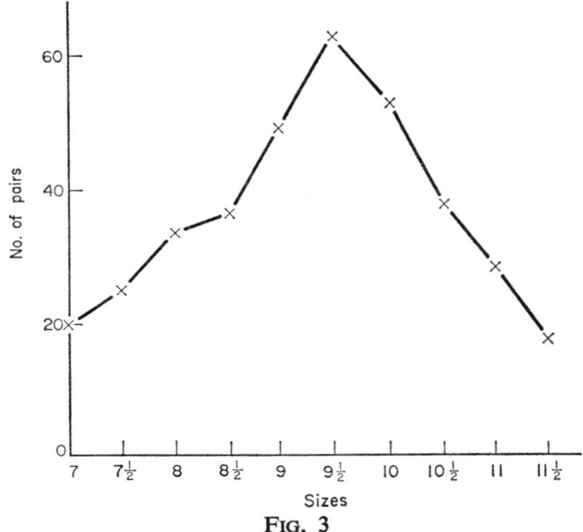

FIG. 3

with which the various sizes are sold. If we join the points by straight lines we have a *frequency polygon*.

It would be quite legitimate to represent the numbers of shoes sold by heavy vertical lines, and indeed this is sometimes the method adopted. Either diagram tells us immediately the size which is in greatest demand, i.e. $9\frac{1}{2}$. This is called the *modal* value of the distribution. It also tells us something about the demand for very small and very large shoes, and, of course, a shoe store manager will base his orders for stock on this sort of information.

Consider now the following table which shows the salaries earned by a group of men belonging to a certain club. For convenience they are placed in order.

£	£	£	£	£
250	500	820	1,100	1,350
280	520	840	1,100	1,400
300	550	860	1,180	1,490
350	620	900	1,200	1,510
360	650	920	1,250	1,600
		.... Median		
400	650	920	1,250	1,650
420	720	950	1,290	1,790
420	740	980 Mean	1,300	1,910
480	780	1,000	1,310	2,100
490	800	1,050	1,320	2,400

Were we to form a scale of values £250, by £1, to £2,400 and plot the frequency, i.e. the number of men corresponding to each salary, the resulting frequency polygon would be quite useless. It would consist of a series of points one unit above the horizontal axis, with an occasional "jump" to a point two units above (e.g. £420 has frequency 2). No pattern would be revealed. If, however, we group the values into equal *class-intervals* of, say £500, we have:

Class interval	*No. of men or Frequency*
£0–499	10
500–999	18
1,000–1,499	15
1,500–1,999	5
2,000–2,499	2

Representing the frequency in each interval by a column of corresponding height, we obtain the diagram shown in Fig. 4. This is called a *histogram*.

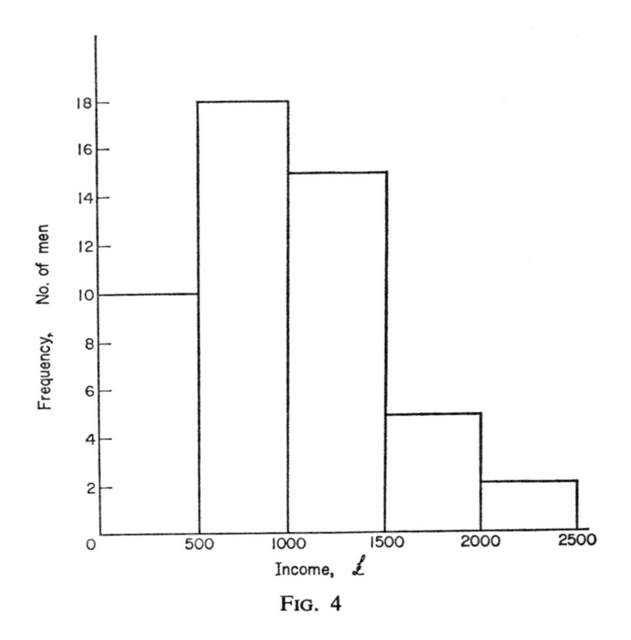

FIG. 4

By this means a pattern is clearly revealed. The *modal* income group here is £500–999.

ANALYSIS OF THE DATA—MODAL, MEAN AND MEDIAN VALUES

So far we have been content to create a picture of our data without making any specific measurements beyond the modal value. The *modal value* is, as the name suggests, the most popular value—the value which occurs with the greatest frequency. It corresponds to the maximum value of the frequency polygon.

Another measurement which is useful in typifying a given set of values or readings is the *mean* or average value. If I know that the group of people I am going to address has an average age of 20 years, then at any rate I have some idea of what they will be like before I actually meet them.

In the last example, the average or mean salary could be obtained by adding together all the separate salaries and dividing the result by 50. You may like to check that this comes to £980. A rather quicker way of obtaining the mean value is to "guess" the answer, add together all the positive and negative *deviations* from this guess or *fictitious mean,* and then divide the total deviation by the number of readings. The average deviation added to the fictitious mean or guessed answer now gives the true arithmetic mean. Thus, suppose that in the example on salaries we take as our "guess" or fictitious mean £1,000. The deviations from this value are shown below. Before making the actual addition we cancel out positive and negative quantities of the same numerical size, e.g. we cancel £−20 and £+20, etc.

Deviations:

£				
−750	−500	−180	100	350
−720 −20	−480	−160	100	400 20
−700	−450	−140	180	490 10
−650	−380 −60	−100	200	510
−640	−350	−80	250	600 100
−600	−350	−80	250	650
−580	−280	−50	290 10	790 20
−580	−260 −10	−20	300	910 50
−520	−220	0	310 10	1,100
−510	−200	50	320	1,400
−580	−510		10	100

Total deviation $= -980$

$$\text{Mean deviation} = \frac{-980}{50} = £-19\tfrac{3}{5}$$

$$\therefore \text{ true mean} = £(1000-19\tfrac{3}{5}) = £980\ 8s.\ 0d.$$

or £980 to the nearest pound.

In the absence of this detailed list of salaries we should still be able to obtain an approximate mean from the histogram in Fig. 4. In such circumstances we should take as average the mid-value in each interval. An approximate mean would then be:

$$\frac{£250\times 10+750\times 18+1{,}250\times 15+1{,}750\times 5+2{,}250\times 2}{50}$$

or £960.

A third kind of number used to typify a set is the *median* value or middle value. In a set of nine values placed in order of magnitude, the median value is the fifth value. In the example discussed above there is an even number of salaries. It so happens that both the twenty-fifth and twenty-sixth salaries are £920, and this is the median salary. If these had been different we could either have quoted both values or the mean of the two values. For a distribution which is perfectly symmetrical the modal, mean and median values all coincide, but for one which is "skew" there is an approximate relation between the three of the form:

$$\text{mode}-\text{median} = 2(\text{median}-\text{mean}).$$

Exercise 9(*b*)

1. Twelve people, asked to guess the weight of a cake to the nearest half-pound, gave the following answers: $3\frac{1}{2}$, 3, 4, $4\frac{1}{2}$, $3\frac{1}{2}$, 3, 5, 4, $3\frac{1}{2}$, 2, $3\frac{1}{2}$, 4 lb. Draw a frequency polygon. What is (a) the modal value, (b) the median value, (c) the mean value?

2. The distribution of the ages of people living in the United Kingdom in 1956 was as shown below. Express the frequencies in units to the nearest 100,000 and draw the histogram.

Years	*No. of people (in thousands)*
0–4	3,863
5–9	4,280
10–14	3,711
15–19	3,304
20–24	3,286
25–29	3,432
30–34	3,691
35–39	3,509
40–44	3,710
45–49	3,743
50–54	3,507
55–59	3,023
60–64	2,536
65–69	2,124
70–74	1,675
75–79	1,149
80–84	605
85 and over	282

(*Monthly Digest of Statistics*, Sept. 1957)

From your approximated frequencies obtain an approximation to the average age of the population.

3. The mean hours of sunshine recorded per day over the years 1921–50 were as follows:

January	1·51
February	2·31
March	3·76
April	5·02
May	6·09
June	6·70
July	5·82
August	5·47
September	4·40
October	3·18
November	1·89
December	1·34

(*Monthly Digest of Statistics*, Sept. 1957)

Represent the data by means of a histogram. What is the expected average number of hours of sunshine per day throughout the year? (Assume each month to be of equal duration.)

4. The number of teachers employed in England and Wales from 1946 to 1956 were as follows:

Year	*No. of teachers employed (thousands)*
1946	190·5
1947	200
1948	211
1949	219
1950	230·8
1951	237·9
1952	243·9
1953	250·7
1954	258·9
1955	265·3
1956	273·9

(*Monthly Digest of Statistics*, Sept. 1957.)

Represent the data by means of a histogram.

5. The weights of 100 boys in a certain form are as follows:

4–6 stones	3
6–8 stones	29
8–10 stones	41
10–12 stones	21
12–14 stones	5
14–16 stones	1

Draw a histogram. State the modal and median values. Calculate the mean value.

6. 30 students carried out experiments to determine Joule's mechanical equivalent of heat. The results, corrected to two significant figures, were as follows:

Value obtained	3·8	3·9	4·0	4·1	4·2	4·3	4·4	4·5	4·6
Frequency	1	1	6	6	7	5	2	1	1

Draw the histogram and determine the mean value of the readings.

7. 15 weighings of the same quantity of a chemical compound were as follows:

13·20, 13·25, 13·28, 13·32, 13·40, 13·29, 13·31,
13·28, 13·35, 13·29, 13·30, 13·29, 13·36, 13·32, 13·30 grams.

State the modal and median values and calculate the mean.

8. The time of ten swings of a given compound pendulum was recorded by 16 boys as follows:

20·4, 20·6, 20·8, 21·0, 21·0, 21·2, 18·8, 20·4,
20·8, 20·6, 21·2, 20·8, 21·0, 20·6, 20·8, 21·0 secs.

What do you think is a reasonable mean value of this set of readings?

9. 15 students performed an acid–alkali titration from given standard solutions. The end-points obtained are given in the following table:

16·30, 16·45, 16·90, 16·00, 16·40, 16·55, 16·45,
17·50, 16·10, 16·25, 16·60, 16·50, 16·35, 16·40,
16·45 ccs.

State the median value and determine the mean value.

10. The following table is a set of examination marks obtained by two unstreamed classes of 32 boys:

7	29	38	44	48	53	58	67
12	31	39	44	48	54	59	69
15	32	39	45	48	54	59	69
18	32	40	45	49	54	60	71
20	34	41	45	50	54	63	74
21	36	41	46	50	55	64	75
23	37	42	47	51	56	64	79
26	37	43	47	53	57	66	87

Arrange these in class-intervals of ten marks 0–9, 10–19, etc., and draw the histogram. Determine the mean or average mark by a quick method.

11. A week later the same group of 64 boys, having gone over their mistakes, sat a similar examination paper. The results are shown below:

12	34	44	52	56	59	65	72
21	35	45	52	57	60	65	73
23	38	46	53	57	60	66	76
26	39	47	54	57	63	67	76
29	40	47	54	58	63	67	78
30	42	49	55	58	64	67	79
32	42	51	56	58	64	68	85
33	43	51	56	59	65	69	89

Arrange, as before, in class-intervals of ten marks. Determine the mean and median values. What differences do you notice between this distribution and the previous one? Comment on these differences.

12. For the set {2, 4, 6, 8, 10} show that the mean of the means of all possible pairs of elements is equal to the mean of the set itself. Does the same apply to the set {2, 4, 8, 16, 32, 64}?

13. The possible results of tossing three coins are as follows:

HHH, HHT, HTH THH, HTT, THT,
TTH, TTT.

Thus we have:

No. of heads	0	1	2	3
Frequency	1	3	3	1

Draw the histogram. Now construct tables showing the possible results of tossing (a) 4 coins, (b) 5 coins. Draw the histograms. Do you notice a significant pattern in these results?

14. In a set of readings the numbers $x_1, x_2, x_3 \ldots x_n$ occur with frequencies $f_1, f_2, f_3 \ldots f_n$ respectively. Let M be the true mean and A a "guessed" or fictitious mean. Show that

$$M = A + \frac{f_1(x_1 - A) + f_2(x_2 - A) \ldots + f_n(x_n - A)}{f_1 + f_2 \ldots + f_n}$$

DISPERSION OR "SPREAD"

Although mean, modal and median values help to typify a set of values, they certainly do not give a complete picture. In the previous section we suggested that if one knew that the audience that one was about to address had an average age of 20 years, this would help to give a "picture" of the group. There are, however, circumstances in which the information might be quite valueless. For example, the audience might consist of 30 people whose ages were as follows:

Age	*No. of people*	
19 years	10	
20 years	10	mean age 20 yrs.
21 years	10	

On the other hand it might consist of the following:

Age	*No. of people*	
10 years	2	
15 years	6	
20 years	14	mean age 20 yrs.
25 years	6	
30 years	2	

The two audiences would be quite different! The picture is more complete if we know the range. In the first case the range is 2 yrs, in the second it is 20 yrs.

To take another example: there is clearly more merit in obtaining the strength of the Earth's magnetic field as ·18 gauss as the average of the readings ·17, ·18, ·19 gauss (range of error = ·02 gauss), than by obtaining an equally good result from the widely differing readings ·12, ·08, ·34 gauss with a range of error 13 times as great!

But again, the range by itself gives us no idea how closely the values "cluster" round the mean. Consider the two distributions given below in Fig. 5.

Each distribution consists of 40 values or readings; each has a range of 25 and a mean of 12·5. In (i), however, the values cluster closely round the mean, while in (ii) they are far more widely *dispersed* or *scattered*. This dispersion can be measured in several

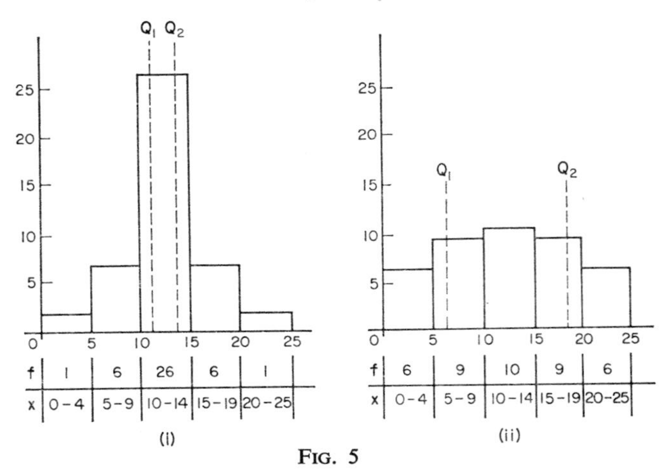

FIG. 5

ways. We may calculate the *mean deviation*, the *standard deviation*, or the *semi-interquartile range*. In this book we shall deal only with the latter, although reference is made to the first two in Exercise 9(c). The median divides the distribution into two equal parts in each of which we have equal numbers of observations. The quartiles divide each of these into two halves: thus, between the two quartiles where $x = Q_1$, $x = Q_2$ respectively, there lies the middle half of the observations. The *semi-interquartile range* is $\frac{1}{2}(Q_2 - Q_1)$, and this clearly gives a better idea of the dispersion than a simple statement of the range.

Exercise 9(*c*)

1. In Question 1, Exercise 9(b), determine the semi-interquartile range of the estimated weights of the cake.

2. In Question 5, Exercise 9(b), mark in the approximate positions of the quartiles on your histogram.

3. In Question 6, Exercise 9(b), state the semi-interquartile range of the experimental results given.

4. In Question 8, Exercise 9(b), state the semi-interquartile range of the experimental readings given.

5. In Question 10, Exercise 9(b), determine the range and the semi-interquartile range of the set of examination marks given.

6. Determine the range and the semi-interquartile range for the set of examination marks given in Question 11, Exercise 9(b).

7. Estimate approximately the semi-interquartile ranges in cases (i) and (ii) of Fig. 5.

8. *The mean deviation* of a set of observations about its mean value is defined as

$$\frac{f_1|d_1|+f_2|d_2|+f_3|d_3|\ldots f_n|d_n|}{f_1+f_2+f_3\ldots f_n},$$

where $|d_1|$ is simply the numerical size of the deviation of the value x_1 from the mean, irrespective of sign. (e.g. if $M = 10$, $x_1 = 12$, $x_2 = 6$ then $d_1 = +2$ and $d_2 = -4$, but $|d_1| = 2$ and $|d_2| = 4$.) f_n is the frequency with which the value x_n occurs in the distribution. Use this formula to determine the mean deviations of the distributions (i) and (ii) of Fig. 5.

9. Determine the mean deviation of the set of experimental results given in Question 9, Exercise 9(b).

10. Determine the mean deviation of the set of weighings given in Question 7, Exercise 9(b).

11. Determine the mean deviation of the set of results given in Question 6, Exercise 9(b).

12. *The standard deviation* σ of a set of observations about its mean value is defined as

$$\sqrt{\left(\frac{f_1 d_1^2+f_2 d_2^2\ldots f_n d_n^2}{f_1+f_2\ldots f_n}\right)}.$$

Use the formula to determine the standard deviation of the distributions (i) and (ii) of Fig. 5.

13. Calculate the standard deviation of the set of experimental results given in Question 9, Exercise 9(b).

14. Calculate the standard deviation of the set of weighings given in Question 7, Exercise 9(b).

15. Calculate the standard deviation of the set of results given in Question 6, Exercise 9(b).

16. Show that if σ is the standard deviation about the true mean M, and S is the "standard deviation" about some fictitious mean A, then $\sigma^2 = s^2 - (M - A)^2$.

THE NORMAL DISTRIBUTION CURVE

Certain features such as height, weight, shoe size and intelligence are distributed in a very symmetrical way over the whole population. A very large number of results in a given experiment

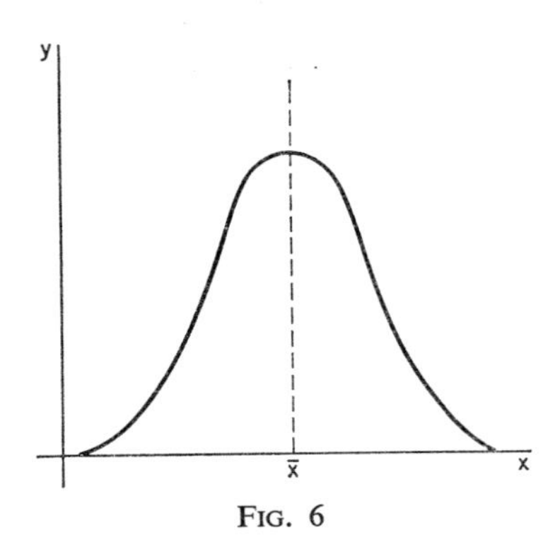

FIG. 6

usually shows the same sort of pattern. If the number of readings or observations is extremely large, the class-intervals may be diminished until the profile of the histogram becomes a curve. The frequency curve for a large distribution is shown in Fig. 6.

PROBABILITY AND CHANCE

Let us take another look at the group of salaries illustrated by the histogram in Fig. 4. There we had a group of 50 men. What is the chance (or colloquially "what are the odds") that if we picked one of these men at random he would be in the highest income group £2,000–2,499? Now only 2 out of the 50 men are in this group, so that for every 50 random choices we shall only make the required selection twice, i.e. out of 25 choices we are likely to include one of these two men. This will not necessarily be the case, but spread over a large number of trials one of these men will turn up in the first 25 choices. We say that the *chance* or *probability* that a man picked at random will be in the highest income group is 1 in 25, or $\frac{1}{25}$. Notice also that the area representing that part of the histogram which relates to this income group is also $\frac{1}{25}$ of the whole area of the histogram. The probability that an event will take place is always measured as a fraction between 0 and 1. If there is only one man in the club and he is in the required income group, we are certain to pick him at the first attempt, i.e. *a probability of* $\frac{1}{1}$ or 1 *represents absolute certainty*. On the other hand, if there is only one man in an infinite group of men, we are unlikely ever to find him, so that $\frac{1}{\text{infinity}}$, or $\frac{1}{\infty}$, or 0, *represents the impossibility of an event occurring*. Hence we define the probability that an event will happen as:

$$\frac{\text{the number of ways in which an event can happen}}{\text{the total number of ways in which the event can either happen or fail to happen.}}$$

Thus, if the event can happen in p ways and fail to happen in q ways, then the probability that it will happen is $\frac{p}{p+q}$, and the probability that it will fail to happen is $\frac{q}{p+q}$. It is *certain* that it will either happen or fail to happen, and so *certainty* is represented by $\frac{p}{p+q}+\frac{q}{p+q} = \frac{p+q}{p+q} = 1$.

Example (i). What is the probability that with one throw of the dice one may (a) score 3, (b) score at least 3? What is the probability that one may throw a 2 and on the next throw score 3?

(a) No. of ways of throwing a 3 = 1.
No. of ways of throwing any number = 6.
$\therefore$ probability of throwing a 3 = $\frac{1}{6}$.
(b) No. of ways of scoring at least 3 = 4(3, 4, 5 or 6).
No. of ways of scoring = 6 (1, 2, 3, 4, 5 or 6).
$\therefore$ probability of throwing at least 3 = $\frac{4}{6} = \frac{2}{3}$.
(c) No. of ways of throwing a 2 and then a 3 = 1.
Total no. of ways of throwing two numbers in succession = $6 \times 6 = 36$.

$$\begin{bmatrix} 1, 1 & 2, 1 & 3, 1 & 4, 1 & 5, 1 & 6, 1 \\ 1, 1 & 2, 2 & 3, 2 & 4, 2 & 5, 2 & 6, 2 \\ 1, 3 & 2, 3 & 3, 3 & 4, 3 & 5, 3 & 6, 3 \\ 1, 4 & 2, 4 & 3, 4 & 4, 4 & 5, 4 & 6, 4 \\ 1, 5 & 2, 5 & 3, 5 & 4, 5 & 5, 5 & 6, 5 \\ 1, 6 & 2, 6 & 3, 6 & 4, 6 & 5, 6 & 6, 6 \end{bmatrix}$$

Hence, the probability required = $\frac{1}{36}$.
Note that the probability of throwing a 2 is $\frac{1}{6}$, the probability of throwing a 3 is $\frac{1}{6}$, and the probability of throwing these in succession is $\frac{1}{6} \times \frac{1}{6} = \frac{1}{36}$.

In general, if the probability of an event X happening is x, and the probability of an event Y happening is y, then the probability that an event X will be followed by an event Y is xy.

Example (ii). Given a pack of 52 playing cards, what is the probability of picking at random (a) a club card, (b) a picture card, (c) two picture cards in succession, (d) a club picture card?

(a) No. of ways of picking a club card = 13.
No. of ways of picking any card = 52.
$\therefore$ probability of picking a club card at random = $\frac{13}{52} = \frac{1}{4}$.
(b) No. of ways of picking a picture card = 12 (King, Queen or Jack from each of 4 suits).
No. of ways of picking a card = 52.
$\therefore$ probability of picking a picture card = $\frac{12}{52} = \frac{3}{13}$.

(c) No. of ways of picking a second picture card = 11 (one has been picked already).
No. of ways of picking any card = 51.
$\therefore$ probability of second card being a picture card = $\frac{11}{51}$.
$\therefore$ probability of two picture cards in succession =
$\frac{3}{13} \times \frac{11}{51} = \frac{11}{221}$.

(d) No. of ways of picking a club picture card = 3.
No. of ways of picking any card = 52.
$\therefore$ probability of picking a club picture card = $\frac{3}{52}$.

Note. If $\mathscr{E}$ is the set of all cards, P the set of picture cards, and C the set of club cards, then the probability $= \dfrac{n(P \cap C)}{n\mathscr{E}}$.

Exercise 9(*d*)

1. In Exercise 9(b), Question 5, find the probability that a student picked at random will weigh:

(a) between 12 and 14 stones,
(b) less than 8 stones.

If two students are picked at random, what is the probability that:

(c) they will both weigh less than 8 stones,
(d) they will both be in the lowest class-interval,
(e) they will both be in the highest class-interval,
(f) they will both be in the modal class-interval,
(g) they will both be in the same class-interval?

2. In Exercise 9(b) Question 6, find the probability that a student picked at random obtained a result correct to 2 significant figures of (a) 4·0, (b) 4·4. What is the probability that 2 students picked at random (c) both obtained 4·1 as their answer, (d) both obtained between 4·0 and 4·2 inclusive as their results?

3. In Exercise 9(b), Question 10, find the probability that a boy picked at random scored:

(a) between 50 and 59 marks,
(b) more than 50%,
(c) less than 40%.

If the pass mark was 40%, find the probability that 2 boys picked at random both failed.

4. In Exercise 9(b), Question 11, find the probability that 2 boys picked at random both failed in the second test.

What is the probability that 3 boys picked at random all scored 43% or less in the second test?

5. In Exercise 9(b), Question 13, find the probability that when 3 coins are tossed (a) no heads are obtained, (b) 2 heads are obtained, (c) at least 2 heads are obtained.

If 4 coins are tossed find the probability that (d) 2 heads are obtained, (e) 2 heads are not obtained.

6. Find the chance or probability of dealing from a full, well shuffled pack of cards:

(a) a club card and then a diamond,
(b) 2 heart cards in succession,
(c) 3 aces in succession,
(d) the King, Queen and Jack of spades in succession,
(e) a King, a Queen and a Jack of any suit in succession,
(f) a King, a Queen and a Jack of any suit and in any order.

7. On the average a boy gets three-quarters of his multiplication "sums" right and half of his division "sums". What is the probability that he will evaluate $\dfrac{2{\cdot}7 \times 4{\cdot}5}{7{\cdot}2}$ correctly?

8. 81 sixth-form pupils took A-level examinations in mathematics, physics and chemistry. 58 passed in mathematics and physics, 51 passed in physics and chemistry, 56 passed in mathematics and chemistry, 2 passed in mathematics only, 3 passed in chemistry only and 65 passed in physics. What is the chance of picking at random a pupil who obtained (a) 3 passes, (b) 1 pass?

9. If x, y can take only values from the set $\{1, 2, 3, 4\}$, find the probability that one of the possible ordered pairs (x, y), picked at random, (a) satisfies the inequality $y > x$, (b) satisfies the equation $y = x$.

10. If any element of the set $\{1, 2, 3, 4, 5\}$ is multiplied by any element of the same set, find the probability that the product is greater than 9. What is the probability if we decide to exclude the cases where an element is multiplied by itself?

10
TOPOLOGY

MUCH of our study of geometry is concerned with lengths and angles, areas and volumes, and the shapes of various figures.

Naturally we are interested in figures which are congruent, those which have equal areas, and those which have similar shapes. We study the relationships between figures which are the projections of each other and those which may be linked by, say, a matrix transformation.

It might be thought that a geometry which concerned itself with none of these things was a very colourless and impracticable business, but in fact this is not the case at all, for when one solid or surface is twisted or stretched without fracture into quite a different form, certain properties remain unaltered. Indeed, the study of these properties has, in very recent times, grown into one of the most important branches of mathematics, and one which now invades many other branches of the subject as well. Furthermore, some of its problems and discoveries are of the greatest significance and fascination.

ORANGES AND DOUGHNUTS

Given a piece of modelling clay and a certain degree of artistic skill, I can form the shape of an orange. I can then remould this into a whole variety of other shapes; for example, an acorn, a pancake, a bolt, a bowler hat, a basin, a shoe horn, a crochet hook or a toy soldier. Every one of these shapes is achieved by a *continuous deformation* of the material; I neither break it, nor do I make holes in it. Each shape can be imagined as flowing in a continuous, smooth fashion into the next.

Of course, if the density of the material remains constant, each model will have the same volume, but for the present purposes this does not interest us, and in fact we may assume that under compression the density does alter. Even so certain features remain invariant.

The main invariant is that if, on any one of these objects, we draw a closed curve, perhaps a small circle or loop, then this circle can be shrunk continuously until it becomes a single point on the surface.

Closely related to this is the property that every one of these objects can be cut into two portions by a *single* knife cut. We say that the surfaces of these objects are *topologically equivalent*.

Suppose now that I break into the modelling clay, forming a hole. I can now make a rather indigestible doughnut. Without making any further holes or breaks I can now deform the clay into a whole variety of other shapes. For example I can make a hexagonal nut, a wedding ring, a needle, a teacup, a jug, a shopping basket, a filter funnel or a smoker's pipe. (For the teacup the hole is in the handle, not in the cup itself. A beaker would *not* be a member of this set.)

With respect to the criteria mentioned above, these objects are alike, but they are essentially different from the first group. In the first place, it is possible to draw a circle, a loop, or a closed curve on the surface of any one of these objects which cannot be shrunk continuously to a point. Furthermore, in each case it is possible to make a single cut which does not make the object fall into two parts. These points are illustrated in Fig. 1 below.

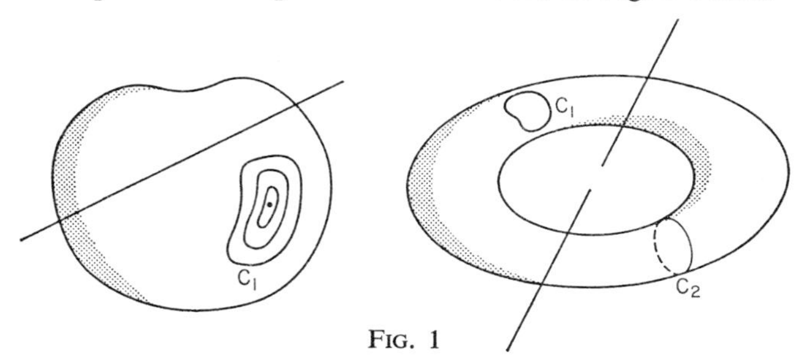

FIG. 1

Closed curves of the type C_1 can be shrunk to a point without leaving the surface; C_2 cannot be so diminished and no curves of this type can be drawn on the surface of an orange.

If I now make *two* holes in the surface of the clay I can clearly form a whole class of new topologically equivalent surfaces. A shirt button, a watering can and a pair of pyjama trousers would all fall into this category, and the latter may require *three* tears before it falls into two pieces of cloth.

Of an even more complicated order we have objects having many holes, of which a flute, a ladder, a spoked wheel and a shoe (with lace holes) are examples. All the surfaces mentioned above are *connected.* A fly, placed at a given point on any one of them, can always reach any other point on the surface by *walking*. However, we say that the orange and its topological equivalents are *simply connected*—all the other groups mentioned are *not* simply connected. Some writers, for topological purposes, would describe the first group as *spheres*, since they are all topologically equivalent to the sphere. The second class is called the *torus* and subsequent classes are described as two-holed or *two-fold* toruses, three-fold toruses . . . n-fold toruses, etc.

Exercise 10(a)

1. Place the following objects in the four categories A, B, C, D:

A topologically equivalent to the sphere,
B topologically equivalent to the torus,
C topologically equivalent to the two-fold torus,
D n-fold toruses where $n > 2$.

A cube
A conker
A tap washer

A pair of spectacle frames
A sausage
A plant pot
A bachelor's button
A five mile length of copper wire
A knitting needle
A penny whistle
A cylinder head gasket
A loaf of bread
A teapot
A glove
A jacket

An electric light bulb
A book
A front door (with keyhole and letter box)
A peanut
A test-tube
A blow pipe
A figure seven
A ladder
A letter R
A letter B
A balloon
A corn plaster
A Wellington boot
A Yale key
A skipping rope

2. Give five other examples of common shapes or objects which have simply-connected surfaces.

3. Give five other examples of common shapes or objects which are topologically equivalent to the torus.

TWISTED SURFACES

Topology has been called "the study of figures which survive twisting and stretching"; it has also been called "rubber sheet geometry".

Suppose we have a rectangular sheet of paper or rubber, *PQRS*. If we join (or identify) the edges $\overrightarrow{PS}$ and $\overrightarrow{QR}$ we clearly form a *cylinder*. The first thing to notice about this is that it has an inside and an outside. An insect, unable to leave the surface and forbidden to cross the edge of the paper, would remain forever either on the inside or the outside of the cylinder.

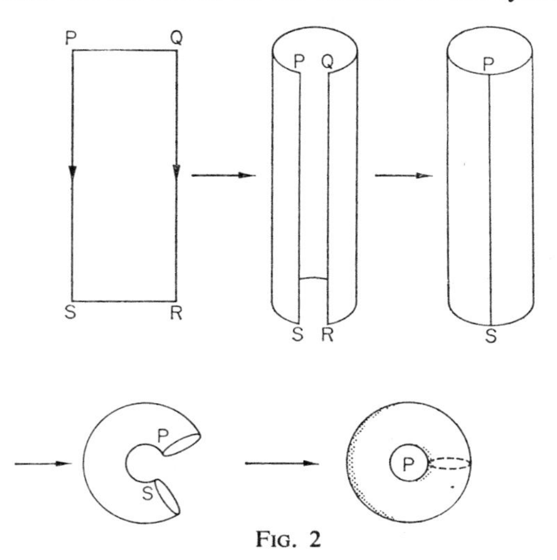

FIG. 2

The second thing to notice is that if we now join together (or identify) the circular edges through *P* and *S* respectively, we form a torus (see Fig. 2).

If, however, instead of joining the edges PS and QR, we first give one end of the paper a half-twist and then join the edges (i.e. we join the edges $\overrightarrow{PS}$ and $\overrightarrow{RQ}$), we obtain what is called a *Möbius strip*. This surface has some extremely interesting properties.

In the first place it is *one-sided*. If we mark any two points A, B on the surface (see Fig. 3), it is now *always* possible for our insect to crawl from A to B without crossing the edge of the strip.

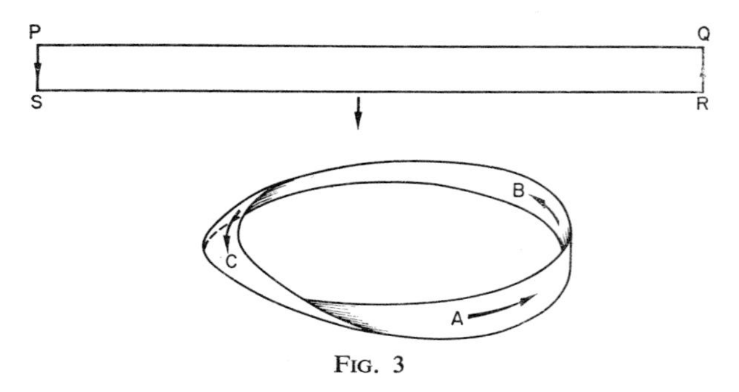

FIG. 3

Secondly, if we cut along the middle of the strip until we have completed one circuit, the strip falls into a loop half as wide, twice as long, and having *four half-twists* in it!

The property of one-sidedness means that if an endless transmission belt were formed as a Möbius strip, wear would be spread evenly over *both* sides of the rectangular strip $PQRS$ from which it was made. In fact this idea has been patented and put to commercial use.

Another interesting feature of the Möbius strip is that it can be distorted into a shopping bag—the only bag which has either an inside or an outside but not both! You can imagine this if you think of the region C (Fig. 3) being tremendously stretched, while the edges between which C lies remain unstretched. Indeed, the type of sling used to immobilize a broken collar bone can be formed as what is essentially a Möbius strip.

A practical illustration of the use of another type of Möbius strip occurs in modern civil engineering. The simple figure-of-

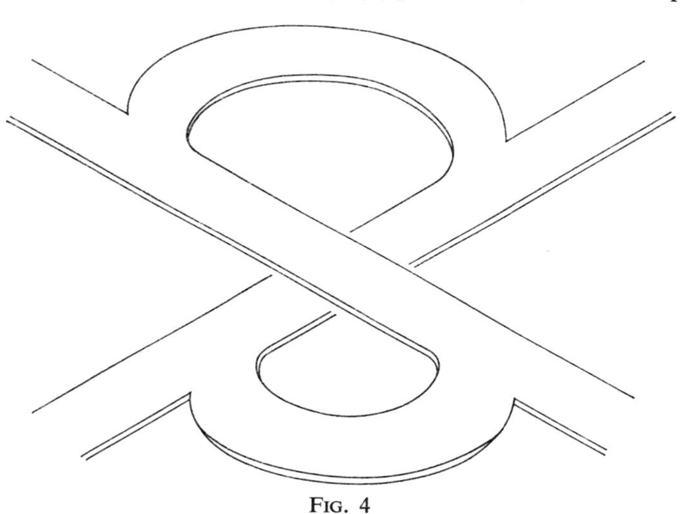

FIG. 4

eight fly-over road link shown in Fig. 4 is actually a Möbius strip containing *two* half-twists. You may like to make a model and check that this is so.

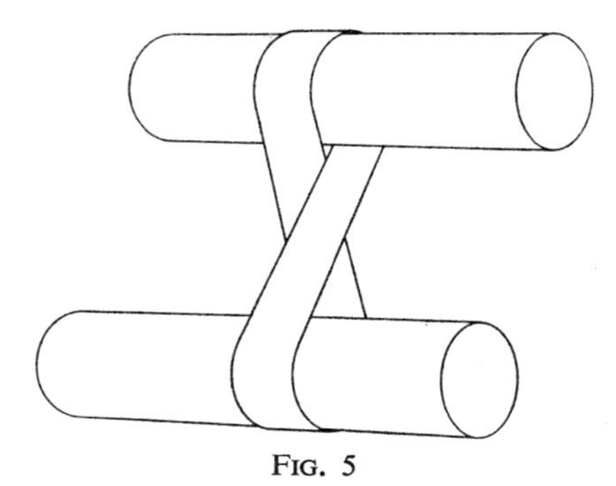

FIG. 5

Exactly the same strip is used in crossed-belt transmissions as shown in Fig. 5.

If we now cut the Möbius strip with two half-twists along a line parallel to its edge, we find that it falls into *two separate interlocking bands*, each containing two half-twists!

By now you may have decided that these curious properties are worth investigating. Exercise 10(b) contains a number of suggestions for doing this.

Exercise 10(*b*)

1. Take a strip of paper marked on both sides with a continuous pattern of oriented circles as shown in Fig. 6.

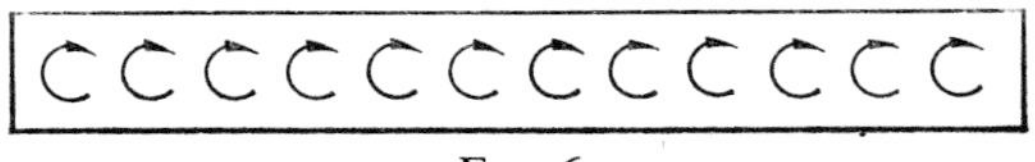

FIG. 6

Form this into (a) a cylinder, (b) a Möbius band. Can you find a point on either band at which the *sense* of rotation of two adjacent circles suddenly changes? You should be able to find such a point on the Möbius band. When this occurs we say that the surface is *non-orientable.* The *cylinder* has an *orientable* surface.

2. Construct a Möbius band which contains *two* half-twists. By marking it and cutting it along the middle investigate its properties.

(a) Is this surface one-sided or two-sided?

(b) Is the surface orientable or non-orientable?

(c) Cut the strip along its centre line. Describe the results of this operation in detail.

In (c) you should find that the strip falls into two separate interlocking bands, each containing the same number of half-twists. How many?

(d) Cut each of these interlocking bands along its centre line. How many strips do you obtain? How do they all interlock? How many half-twists does each strip contain?

3. Form a Möbius band containing *three* half-twists.

(a) Is this surface one-sided or two-sided?
(b) Is this surface orientable?
(c) Cut the strip along its centre line. Describe the result in detail.

4. Form a Möbius band containing *four* half-twists.

(a) Is this surface one-sided?
(b) Is it orientable?
(c) What happens if you cut it along its centre line? Describe the result in detail.

5. Without making any more bands, what would you expect of one containing (a) an odd number of half-twists, (b) an even number of half-twists?

6.

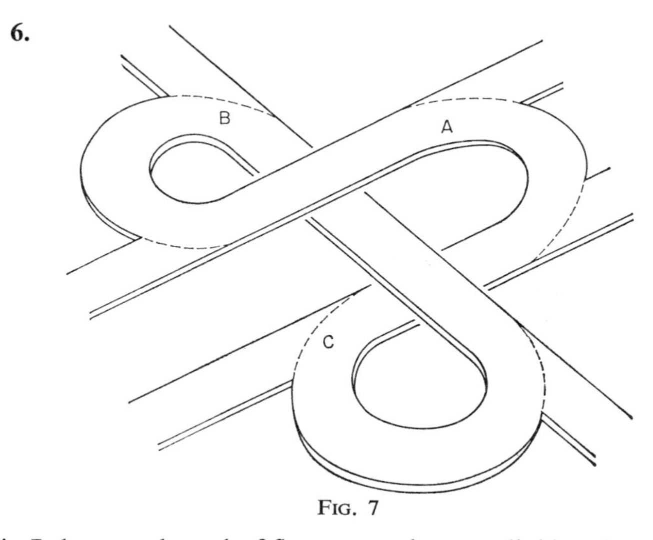

FIG. 7

Fig 7 shows a clover leaf fly-over road system linking three main highways *A*, *B*, *C*, built on different levels. Make a paper model of this linkage. It is essentially a Möbius band. How many half-twists does it contain? Is it an orientable surface? Would you *expect* it to be orientable or non-orientable? Could such a linkage be made for *four* highways situated on different levels? How many half-twists would it contain? What would you call it?

7. Cut a long, narrow strip of paper and tie it as you would tie your own tie. Hold the ends AB and $A'B'$ (see Fig. 8) together

FIG. 8

so that A falls on A' and B on B'. You will find that you have an endless band with a knot in it. Now, *without twisting either end*, carefully disentangle the knot until you have a band which does not contain a knot. (A can still be placed on A' and B on B' without further twists.) How many half-twists does this band contain? (You should find that it contains only one half-twist, i.e. the tie is a knotted Möbius band.)

FACES, EDGES AND CORNERS

The great Swiss mathematician Euler discovered a most important formula connecting the number of faces (F), edges (E), and corners (or vertices) (C) possessed by a solid body. See whether you can discover this relationship for yourself.

Exercise 10(*c*)

1. Complete the following table.

TABLE 1

Figure (*Polyhedron*)	*No. of faces—F*	*No. of corners—C*	*No. of edges—E*
(a) Cube 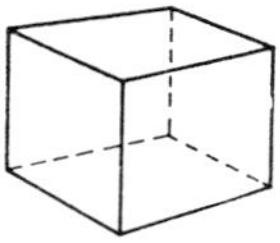			
(b) Triangular prism 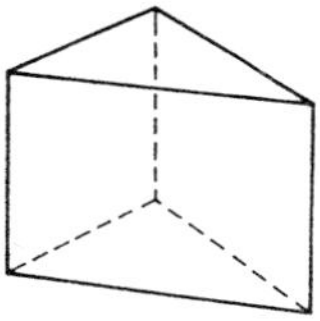			
(c) Pentagonal prism 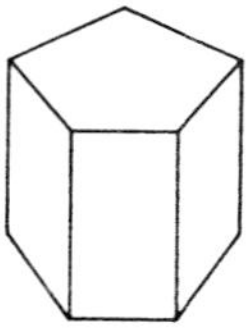			
(d) Hexagonal prism 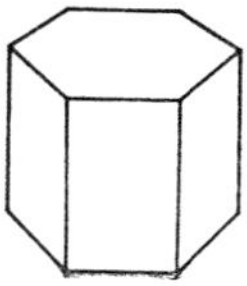			

Figure	*No. of faces—F*	*No. of corners—C*	*No. of edges—E*
(e) Tetrahedron 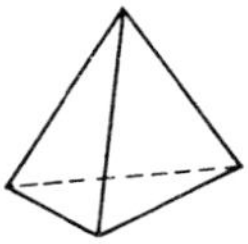			
(f) Square pyramid 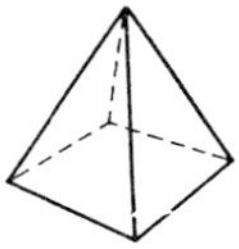			
(g) Pentagonal pyramid 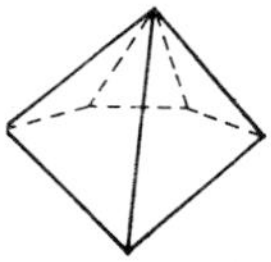			
(h) Prism with pyramidal ends 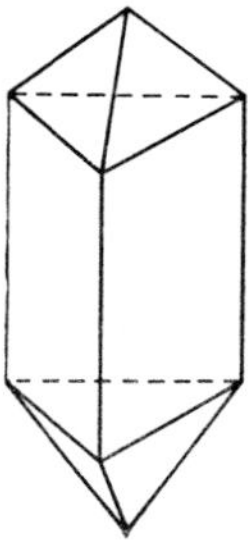	9	8	15

2. The surface of a sphere, or any closed (simply connected) surface, is divided into F regions by means of E arcs meeting at C corners or vertices, so that at least two edges meet at any one

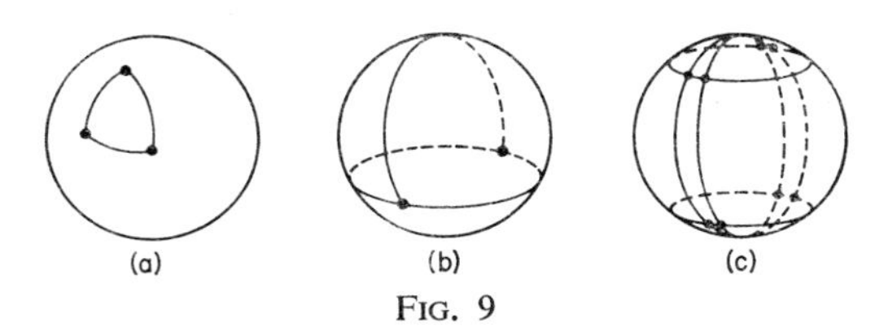

FIG. 9

vertex. Fig. 9 shows three cases. Find the value of $F+C-E$ in each case and investigate whether the same result applies to other cases of your own devising.

3. Fig. 10 shows the result of boring a triangular hole through Fig. (h) of Table 1. Determine the value of $F+C-E$.

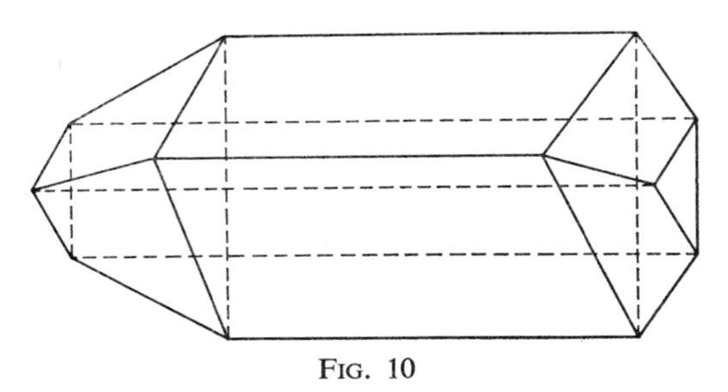

FIG. 10

What is the *topological* name of this shape?

4. In forming the torus as shown in Fig. 2, we identified two pairs of edges of the original rectangle (i.e. four edges became two edges), and we identified all four corners of the original rectangle leaving one vertex P (or S). Find the value of $F+C-E$ for the torus.

5.

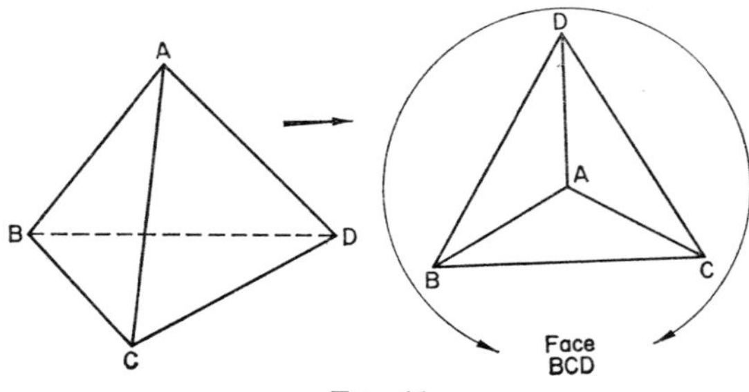

FIG. 11

The surface of the tetrahedron $ABCD$ is deformed into a topologically equivalent surface in the following manner: the edges AB, AC, AD are diminished and the face BCD is greatly expanded until the points A, B, C, D lie on a sphere. (The face BCD is now the whole surface of the sphere, except for the region contained within the arcs BD, DC, CB.) Indeed, we can now imagine this sphere increasing in radius until, its radius infinite, the figure becomes a *plane* triangle BDC *containing the point* A and *surrounded by* the original face BDC, now infinite in extent! (See Fig. 11.) Note that in the transformed case $F = 4$ (one of the faces is infinite and surrounds the figure), $C = 4$, $E = 6$ and $F+C-E = 2$.

Now remove the edge BC. What is the value of $F+C-E$? Next, remove the edges AC, DC. What is the value of $F+C-E$? Instead of removing these edges, add a new edge AE from A to any point E on BC. What is the value of $F+C-E$? Complete the statement of Euler's theorem (commenced at the beginning of Question 2). Can you devise a means of proving the theorem?

If you have worked through Exercise 10(c) you will have discovered that for topologically equivalent surfaces $F+C-E$ is an invariant quantity. For the sphere, and all polyhedra topologically equivalent to it, we have $F+C-E = 2$. That is, for any network on the sphere by which F regions are enclosed by E arcs which meet in C vertices (at least two edges meeting at each vertex), then the expression $F+C-E$ is independent of the network or mode of division of the surface.

The same applies to surfaces topologically equivalent to the torus, except that here $F+C-E = 0$. For the two-fold torus $F+C-E = -2$, and, in general, for the n-fold torus $F+C-E = 2-2n$.

NETWORKS

There is a well known problem in which it is required to run supply lines to three houses, A, B, and C, from three service stations supplying gas (G), electricity (E) and water (W), in such a manner that none of the supply lines intersects. You may like to try this for yourself. If you are unable to solve the problem do not despair—it is insoluble. At least one intersection arises no matter how the houses and stations are situated. One attempt is shown in Fig. 12.

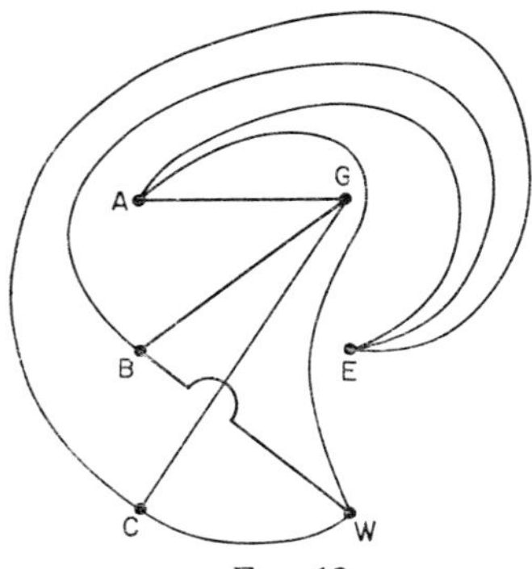

FIG. 12

Of course, we could have *more* than one intersection. The problem of minimizing intersections in networks is of great interest in designing railway lay-outs, in road planning, and in the construction of stamped graphite micro-circuits. In road planning, intersections can sometimes only be avoided by the construction of fly-overs, i.e. the problem is soluble in three dimensions but not in two. By the same token there are topological surfaces and knot problems which exist, or can be unravelled, in four dimensions but not in three!

Exercise 10(*d*)

1. Show that the supply lines from two stations to three houses do not necessarily intersect. Does the same apply if we have three stations and two houses?

2. Show that the least number of intersections possible with four houses and three stations is two.

3. Show that the least number of intersections possible with four stations and four houses is four.

4. Is it true to say that each additional house or station raises the minimum number of intersections by two?

5. Using an inflated inner tube and coloured chalk show that, on the *torus*, A, B and C *can* be linked with each of G, E and W without intersection.

It is not surprising that the problem of networks interested Euler, and that he was the first mathematician to give a satisfactory analysis of the ancient problem of the bridges of Königsberg.

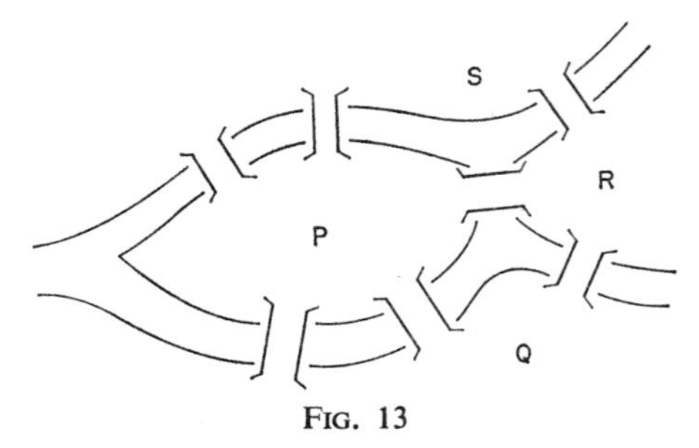

FIG. 13

The position of the bridges is as shown in Fig. 13, and the problem is to cross each bridge *once, and once only*, in the course of a walk. You may like to think about the problem yourself.

Exercise 10(*e*)

1. If you were allowed to remove one bridge in Fig. 13, would it be possible to take a walk during which the remaining bridges were each crossed once, and once only? Which bridge should be removed—or is it sufficient to remove any one bridge?

2. Is it possible, by removing one bridge as in Question 1, to start at any point *A*, cross each of the six bridges once, and once only, and return to *A*? If not, can it be achieved by removing further bridges?

3. Can the problem of the bridges be resolved by *adding* bridges? If so, how many? Where should they be situated?

4. Can the problem of starting at any point *A* and returning thereto while crossing each bridge once, and once only, be solved by the *addition* of bridges? If so, how many? Where must the extra bridges be added?

5. Show that Fig. 13 is *topologically equivalent* to Fig. 14 in which the *regions P, Q, R, S* have shrunk to *points P, Q, R, S*, and the bridges have become lines connecting these points.

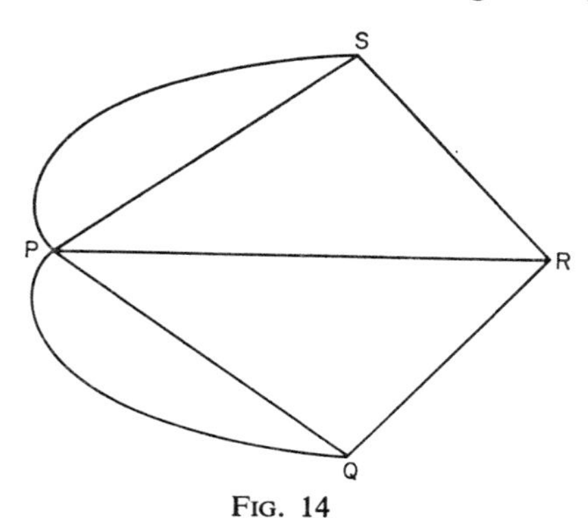

FIG. 14

Hence show that the problem of the bridges in Fig. 13 is *equivalent* to the problem of making a continuous, complete circuit of Fig. 14 without travelling along any line more than once.

If, in Questions 2 and 3, you have resolved the problem by removing or adding a bridge, draw the topologically equivalent network and show that this too may be completely and continuously traversed without duplication. Do you notice anything about a network which can be traversed without duplication?

Euler showed that a simply connected network may be completely traversed in a single journey if, and only if, there are either *no* vertices or *just two* vertices at which an odd number of paths meet. Hence, for example:

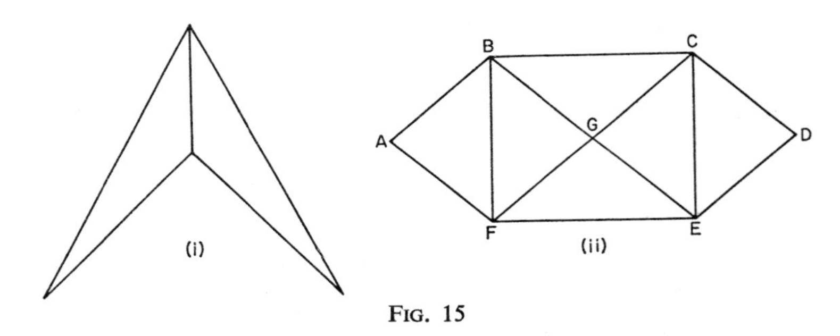

FIG. 15

In Fig. 15(i) there are *two* points at which an *odd* number of paths meet, and it is easy to see how this can be completely traversed, without repetition, in a single journey. In Fig. 15(ii) there are *no* points at which an *odd* number of paths meet, and this can be traversed by the circuit *FABFGBCGECDEF.*

COLOURED BRICKS AND PATCHWORK QUILTS

Suppose that we wish to colour each face of a cube so that any two faces meeting along an edge are painted in different colours. How many colours are required, and, given this number of colours, in how many different ways can the cube be painted? You may like to think about this problem before reading any further.

You will not find it difficult to discover that *three* colours are sufficient, and that if these are, for example, red, blue and yellow, then there is only *one* possible way of painting the cube.

Since the shape of the faces is unimportant, we may, as in Fig. 11, use the methods of "rubber sheet geometry" and imagine the cube distorted into the shape of a sphere. We may then imagine

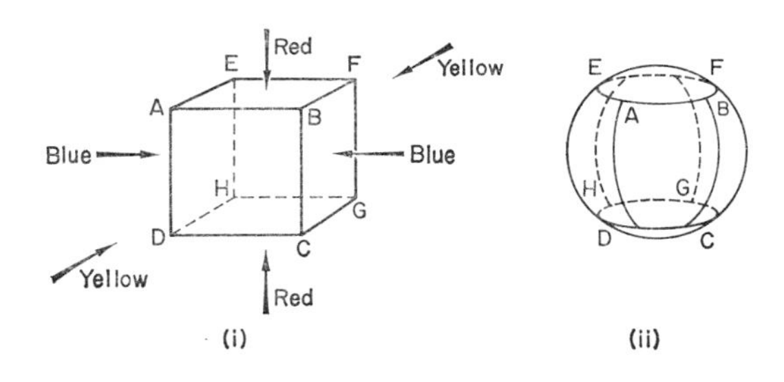

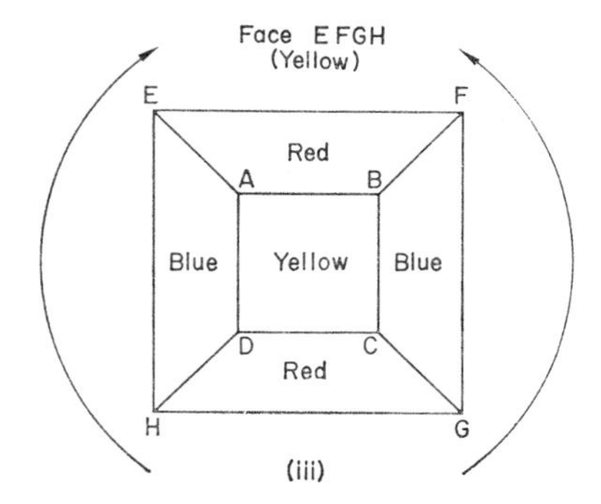

FIG. 16

the radius of this sphere to increase indefinitely by stretching one face of the cube indefinitely. The result is a map on an infinite plane surface as shown in Fig. 16.

This now raises one of the oldest and most interesting questions in topology. How many colours are necessary to colour a map in such a way that regions or countries which meet along an edge, or boundary, or frontier, are coloured differently?

If you look at Fig. 11 again, you will see that, rather surprisingly, *four* colours are necessary in this case.

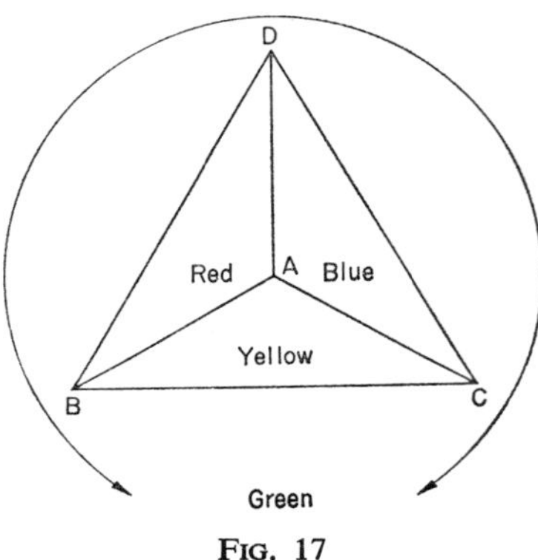

FIG. 17

Furthermore, no one could describe this map as complicated!

Möbius set his students the task of dividing any plane area into five regions so that every two would have a common boundary line. This, if you think about it, is precisely the same problem as that of devising a map which requires five colours to distinguish all its regions. Curiously enough, no such map has ever been discovered, and it remains one of the unsolved problems of mathematics. However, it has been proved that any map with not more than 38 regions or countries can always be coloured with four colours, so if a *five* colour map is ever discovered it will certainly be a very complicated one! In addition, it has also been proved that five colours are sufficient no matter how complicated the map. In other words, although a five colour map has never been discovered, it is *certain* that no six colour map *can* ever be discovered. All this, of course, applies only to maps on plane or spherical surfaces, i.e. for surfaces where $F+C-E=2$.

If, however, we look into the problem of colouring maps on surfaces such as the torus or the Möbius strip of one half-twist, we find that these results do not apply. In fact, although these surfaces in some respects are more complicated than that of the sphere, the problem of colouring maps upon them has been solved completely.

For the torus it has been proved that seven colours are sufficient, and one such map actually exists. This is shown in Fig. 18, where the surface of a torus is shown unfolded back to the rectangle from which it was formed (see Fig. 2).

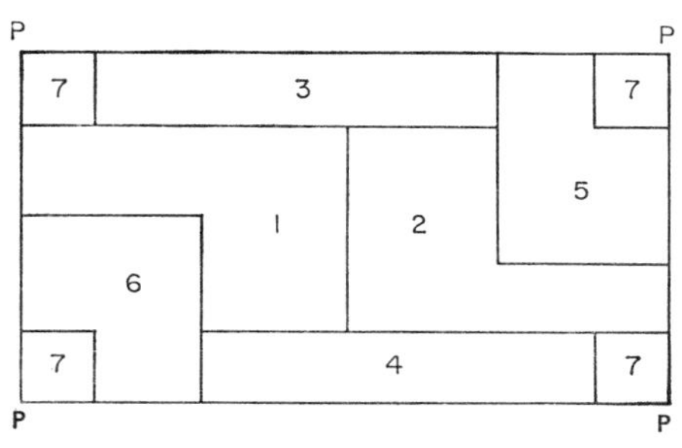

FIG. 18

You should make a model of this surface and satisfy yourself that seven different colours are, in fact, required.

For the Möbius strip it has been proved that six colours are sufficient for any map, and here again one simple, well known case exists (see Exercise 10(f)).

The minimum number of colours required to make *any* map on a given type of surface is called the *chromatic number* of that surface. Like the value of $F+C-E$, to which it is closely related, it is an invariant quantity for all surfaces which are topologically equivalent.

Exercise 10(f)

1. A patchwork quilt is made up of squares of material as shown in Fig. 19. Squares which meet along an edge (not just at

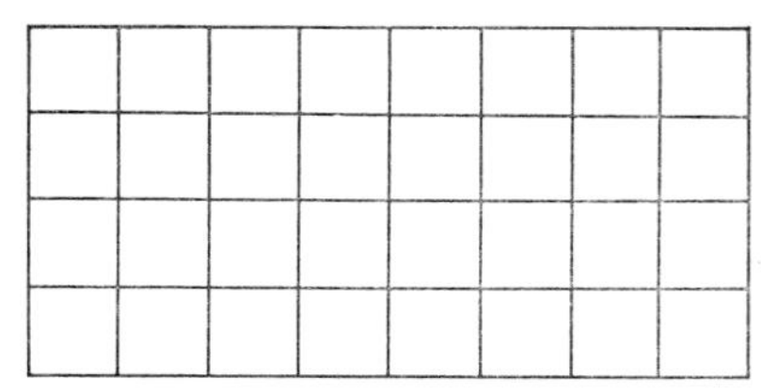

FIG. 19

a point or a vertex) must be of different colours. What is the minimum number of colours required?

2. A patchwork quilt is made in the form of a regular pattern of equilateral triangles.

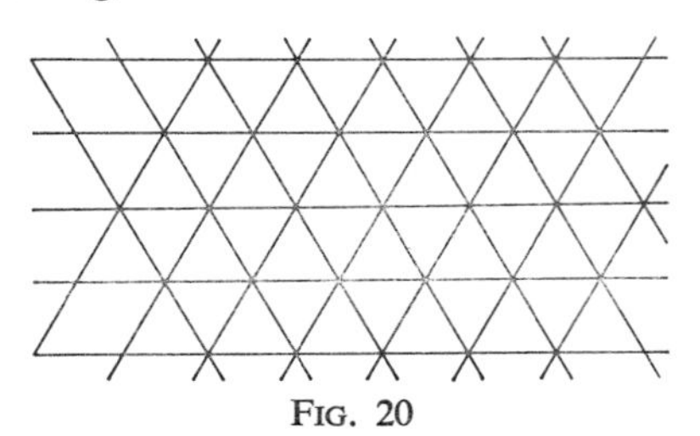

FIG. 20

How many colours are necessary to ensure that adjacent triangles are of different colours?

3. An area of tessellated flooring is to be made in the form of a honeycomb, i.e. the plane is completely covered with regular hexagons. How many colours are necessary to ensure that adjacent hexagons are of different colours?

4. The faces of a tetrahedron (regular triangular pyramid) are to be coloured in such a way that every two adjacent faces are of different colours. How many colours are required? Given the required number of colours, in how many ways can the colouring be done?

5. An octahedron is to be coloured so that no two faces adjacent along an edge have the same colour. How many colours are required, and in how many different ways can the colouring be done?

6. Find the minimum number of colours required to colour distinctively the faces of (a) an icosahedron, (b) a dodecahedron.

7.

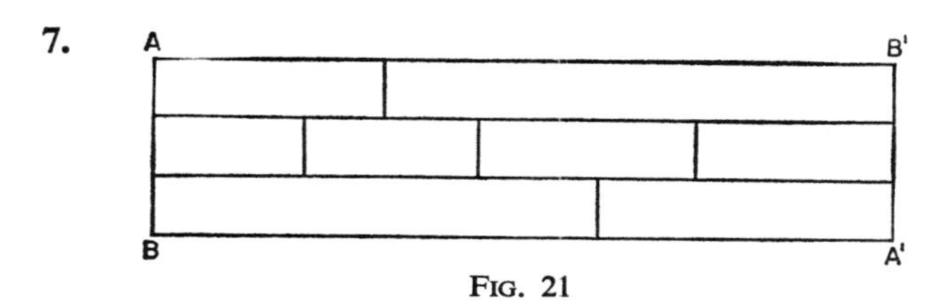

FIG. 21

Mark out both sides of a strip of paper as shown in Fig. 21. Give one end of the paper a half-twist, and by joining A and A', B and B', form a Möbius strip. Investigate the minimum number of colours required for this map (you will have to colour both sides of the paper), and indicate their positions on your own copy of Fig. 21.

ANSWERS

CHAPTER 1

Exercise 1(a)

1. {Sunday, Saturday}
3. {January, June, July}
4. {April, June, September, November}
5. {East Riding, North Riding, West Riding}
6. {1, 2, 3, 4, 5, 6}
7. {Mercury, Venus}
8. {(0, 0) (0, 1) (0, 2) (0, 3) (0, 4) (1, 1) (1, 2) (1, 3) (2, 2)}
9. {6, 7, 8, 9}
10. $\{a, e, i\}$
11. {12, 14, 16, 18}
12. {12, 15, 18}
13. {12, 18}
14. {12, 14, 15, 16, 18}
15. {14, 16}
16. u
17. West.
18. Lane or drive etc.
19. 1968 (or any other Leap Year).
20. John.
21. Maisonette, flat, mansion, villa, or any other dwelling place.
22. Mid-on, fine leg, wicket keeper, gully, slip, etc.
23. Brussels, Berlin, Madrid, Moscow, or any other capital city.
24. Any other breed of dog.
25. Any other odd number.
26. Any other prime number.
27. Any other perfect square. (25, 36, etc.)
28. 64, 128, etc., or any other integral power of 2.
29. $\frac{4}{7}$ or $\frac{5}{6}$ or any other fraction, the sum of whose numerator and denominator is 11.
30. ·142857, or any other decimal approximation to $\frac{1}{7}$.
31. Possible.
32. Possible (but difficult).
33. Possible (but very difficult).
34. Impossible.
35. Possible.
36. Impossible.
37. Possible. There are none. Alternatively the set required is the empty set or $\emptyset$.

38. Possible. (There are 31 different selections of coins but only 25 different sums of money.)
39. Impossible.
40. Impossible.

Exercise 1(*b*)

1. T	**2.** F	**3.** T	**4.** T
5. F	**6.** T	**7.** F	**8.** T
9. T	**10.** T	**11.** F	**12.** T
13. T	**14.** F	**15.** F	({2} is a *set* and can not be an element.)

Exercise 1(*c*)

1. $\{\frac{1}{5}, \frac{2}{5}, \frac{3}{5}, \frac{4}{5}\}$
2. Impossible to list.
3. $\emptyset$
4. Impossible to list.
5. $\emptyset$
6. {105, 112, 119, 126, 133, 1147, 04}
7. Impossible to list.
8. $\emptyset$
9. $\emptyset$
10. $\emptyset$

Exercise 1(*d*)

1. G	**2.** E	**3.** J	**4.** I
5. H	**6.** B	**7.** K	**8.** D
9. F or $\emptyset$	**10.** L		

Exercise 1(*e*)

1. $\in$	**2.** $=$	**3.** $\subset$	**4.** $\notin$
5. $\subset$	**6.** $\subset$	**7.** $\subset$	**8.** $\subset$
9. $\subset$ (If we exclude 2 and 3.)			**10.** $\in$

11. (a) {2, 4, 6, 8}
(b) {6, 7, 8, 9}
(c) {1, 2, 3}
(d) $\emptyset$
(e) $\mathscr{E}$

12. (a) $\{A, B, C\}$ $\{A, B, D\}$ $\{A, C, D\}$ $\{B, C, D\}$
(b) $\{A, B\}$ $\{A, C\}$ $\{A, D\}$ $\{B, C\}$ $\{B, D\}$ $\{C, D\}$
(c) $\{A, C\}$ $\{A, D\}$ $\{B, C\}$ $\{B, D\}$
(d) $\{A, B, C\}$ $\{A, B, D\}$ $\{A, C, D\}$
(e) $\{D, A\}$ $\{D, B\}$ $\{D, C\}$

13. (a) {London, Paris}
(b) {Blackpool, Scarborough} together with subsets of these
(c) {Leeds, Scarborough}

14. (a) 16 (b) 32 (c) 2^n

Exercise 1(*f*)

1. (a) $\{b, c, d\}$ (b) $\{b, c, d\}$
(c) $\{a, c\}$ (d) $\{c, f\}$
(e) $\{c\}$ (f) $\{c\}$

2. J = {January, June, July}
Y = {January, February, May, July}
S = {April, June, September, November}
(a) {January, July} (b) {June}
(c) $\emptyset$ (d) J
(e) Y (f) S
(g) $\emptyset$ (h) $\emptyset$

3. $X = \{2, 3, 4, 5\}$
$Y = \{3, 4, 5, 6\}$
$Z = \{3, 5, 6, 7\}$
(a) $\{3, 4, 5\}$ (b) $\{3, 5\}$
(c) $\{3, 5, 6\}$ (d) X
(e) Y (f) Z
(g) $\{3, 5\}$ (h) $\{3, 5\}$

Exercise 1(*g*)

1. (a) $\{a, b, c, d, e, f\}$ (b) $\{a, b, c, d, e, f\}$
(c) $\{a, b, c, d, f\}$ (d) $\{a, b, c, d, e, f\}$
(e) $\{a, b, c, d, e, f\}$ (f) $\{a, b, c, d, e, f\}$

2. (a) {January, February, May, June, July}
(b) {January, April, June, July, September, November}
(c) {January, February, April, May, June, July, September, November}
(d) $\mathscr{E}$
(e) $\mathscr{E}$
(f) $\mathscr{E}$
(g) {January, February, April, May, June, July, September, November}
(h) {January, February, April, May, June, July, September, November}

3. (a) $\{2, 3, 4, 5, 6\}$
(b) $\{2, 3, 4, 5, 6, 7\}$
(c) $\{3, 4, 5, 6, 7\}$
(d) X
(e) Y
(f) Z
(g) $\{2, 3, 4, 5, 6, 7\}$
(h) $\{2, 3, 4, 5, 6, 7\}$

4. (a) $\{1, 2, 3, 4, 5\}$ (b) $\{1, 2, 3, 4, 6\}$
(c) $\{2, 3, 4, 5, 6\}$ (d) $\mathscr{E}$
(e) $\mathscr{E}$ (f) $\{3, 4\}$
(g) $\{2, 4\}$ (h) $\{4\}$
(i) $\{4\}$ (j) $\{4\}$
(k) $\{2, 3, 4\}$ (l) $\{2, 3, 4, 6\}$
(m) $\{1, 2, 3, 4\}$ (n) $\{2, 4\}$
(o) $\{2, 3, 4\}$ (p) $\{1, 2, 3, 4\}$
(q) A (r) B
(s) C (t) $\mathscr{E}$
(u) $\mathscr{E}$ (v) $\mathscr{E}$
(w) A (x) B
(y) $\emptyset$ (z) A
(k) and (o); (m) and (p) are equal.

5. (a) $\{2\}$ (b) $\{4\}$
(c) $\{7, 8, 9\}$ (d) $\{1, 2, 3, 4\}$
(e) $\emptyset$ (f) $\emptyset$
(g) $\{2, 3, 4, 5, 6, 7, 8, 9\}$ (h) $\{2, 3, 5, 7\}$
(i) $\{1, 4, 9\}$
$\{2, 3, 4\}$; $\emptyset$; $\{2, 4\}$

6. (a) $\{D\}$ (b) $\{B, C, D, I\}$
(c) $\{A, D, E, H\}$ (d) $\{D\}$
(e) $\mathscr{E}$ (f) $\{A, D, E, H\}$
(g) $\{A, D, E, H\}$ (h) $\{A, B, C, D, E, H, I\}$
(i) $\{A, B, C, D, E, H, I\}$
(f) and (g) are equal; (h) and (i) are equal. These expressions are always equal.

Exercise 1(*h*)

1. (a) $\{p, r\}$ (b) $\{m, n, p, r\}$
(c) $\emptyset$ (d) $\mathscr{E}$
(e) $\{r\}$ (f) $\{l, p, r\}$

2. (a) $\{p\}$ (b) $\{m, r\}$
(c) $\{p, m, r\}$ (d) $\emptyset$
(e) $\mathscr{E}$ (f) $\{l, n\}$
(g) $\emptyset$ (h) $\{p, m, r\}$
(i) $\mathscr{E}$ (j) $\emptyset$
(k) $\mathscr{E}$ (l) $\emptyset$
(c) and (h) are equal; (d) and (g) are equal. These expressions are always equal.

3. $J = \{\text{January, June, July}\}$
   $S = \{\text{April, June, September, November}\}$
   (a) {February, March, April, May, August, September, October, November, December}
   (b) {January, February, March, May, July, August, October, December}
   (c) {February, March, May, August, October, December}
   (d) {January, February, March, April, May, July, August, September, October, November, December}
   (e) {February, March, May, August, October, December}
   (f) {January, February, March, April, May, July, August, September, October, November, December}
   (c) and (e) are equal; (d) and (f) are equal.

4. The following are equal: (a) and (b); (c) and (d); (g) and (h); (i) and (j).
   (k) M (l) M (m) M (n) M
   (o) M (p) P

Exercise 1(*i*)

1. (a) A (b) A (c) $\emptyset$ (d) $\mathscr{E}$
   (e) A (f) A (g) A (h) A

2. (a) and (b), (c) and (d), (e) and (f), (g) and (h) are equal.

3. (c) $B \cup A$ (d) $B \cap A$

Exercise 1(*k*)

1. 7
2. 2; 3
3. 5; 4
4. 2
5. 48; 10
6. (a) 123 (b) 135 (c) 405
7. 4

CHAPTER 2

Exercise 2(*a*)

1. (a) {5, 6, 7 . . .} (b) {4, 5, 6 . . .}
   (c) {1, 2, 3, 4, 5, 6, 7} (d) {5, 6, 7}
   (e) The set of positive integers.
   (f) $\emptyset$
   (g) The set of positive integers.
   (h) {6, 7, 8, . . .}

2. (a) $\{2\}$
(b) $\emptyset$
(c) $\mathscr{E}$
(d) $\emptyset$
(e) $\{2, 3, 4 \ldots\}$
(f) $\mathscr{E}$
(g) $\emptyset$

3. (a) $\{2, 4, 6, 8\}$
(b) $\{6, 7, 8, 9\}$
(c) $\{1, 2, 3, 4\}$
(d) $\{3\}$
(e) $\emptyset$

4. (a) $\{2\}$
(b) $\{4, 5, 6 \ldots\}$
(c) $\mathscr{E}$
(d) $\{10\}$
(e) $\emptyset$

Exercise 2(*b*)

1. (a) $\{(1, 1)\ (2, 2)\ (3, 3)\ (4, 4)\}$
(b) $\{(1, 2)\ (2, 3)\ (3, 4)\}$
(c) $\{(1, 1)\ (2, 2)\ (3, 3)\ (4, 4)\ (1, 2)\ (1, 3)\ (1, 4)\ (2, 3)\ (2, 4)\ (3, 4)\}$
(d) $\{(2, 1)\ (3, 1)\ (3, 2)\ (4, 1)\ (4, 2)\ (4, 3)\}$
(e) $\{(2, 4)\ (3, 3)\ (4, 2)\}$
(f) $\{(2, 1)\ (2, 2)\ (2, 3)\ (2, 4)\}$
(g) $\{(1, 4)\ (2, 4)\ (3, 4)\ (4, 4)\}$
(h) $\emptyset$
(i) $\{(2, 4)\}$
(j) $\mathscr{E}$; the whole set.

2. (a) $\{(0, 0)\ (1, 2)\ (2, 4)\}$
(b) $\{(0, 1)\ (1, 2)\ (2, 3)\ (3, 4)\}$
(c) $\{(0, 2)\ (1, 3)\ (2, 4)\}$
(d) $\{(0, 4)\ (2, 3)\ (4, 2)\}$
(e) $\{(0, 4)\ (0, 3)\ (0, 2)\ (0, 1)\ (0, 0)\ (1, 3)\ (1, 2)\ (1, 1)\ (1, 0)\ (2, 2)\ (2, 1)\ (2, 0)\ (3, 1)\ (3, 0)\ (4, 0)\}$
(f) $\{(0, 2)\ (0, 3)\ (0, 4)\ (1, 1)\ (1, 2)\ (1, 3)\ (1, 4)\ (2, 0)\ (2, 1)\ (2, 2)\ (2, 3)\ (2, 4)\ (3, 0)\ (3, 1)\ (3, 2)\ (3, 3)\ (3, 4)\ (4, 0)\ (4, 1)\ (4, 2)\ (4, 3)\ (4, 4)\}$
(g) $\{(1, 2)\}$
(h) $\{(2, 4)\}$
(i) $\{(2, 3)\}$
(j) $\{(0, 2)\ (0, 3)\ (0, 4)\ (1, 1)\ (1, 2)\ (1, 3)\ (2, 0)\ (2, 1)\ (2, 2)\ (3, 0)\ (3, 1)\ (4, 0)\}$

3. (a) {2}

(b) {3}

(c) {2, 3}

(d) ∅

(e) {0, 4}

(f) {(2, 0) (2, 1) (2, 2) (2, 3) (2, 4) (0, 3) (1, 3) (3, 3) (4, 3)}

(g) {(2, 3)}

(h) {(2, 2)}

(i) {(0, 0) (0, 1) (1, 0) (1, 1)}

(j) {(0, 0) (0, 1) (1, 0) (3, 4) (4, 3) (4, 4)}

4.

(a) (b)

(c) (d)

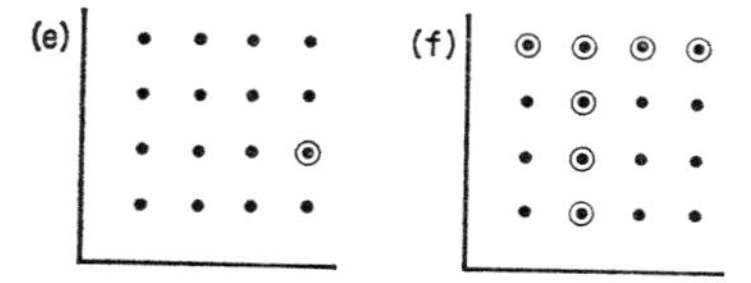

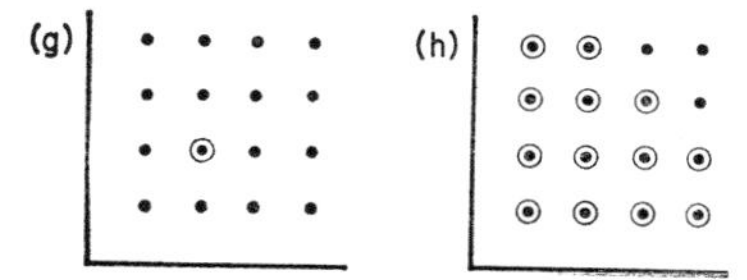

FIG.

Exercise 2(*c*)

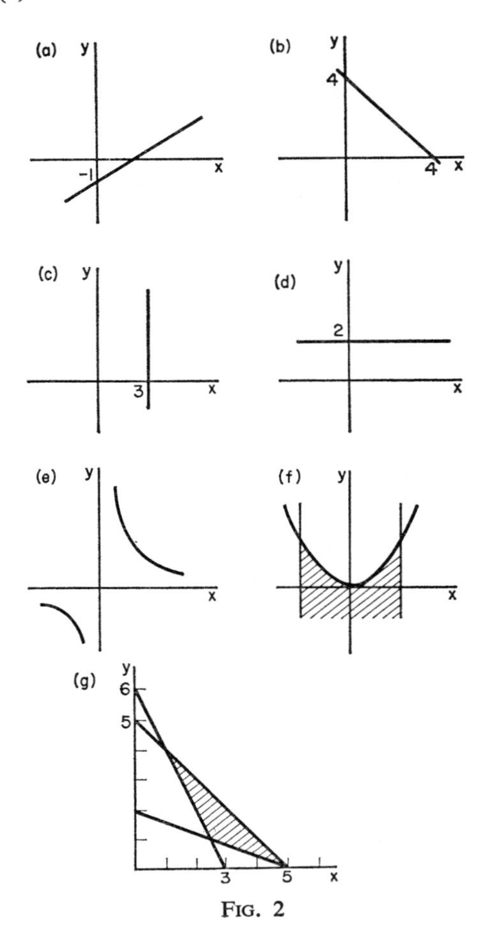

FIG. 2

3. and **4.**

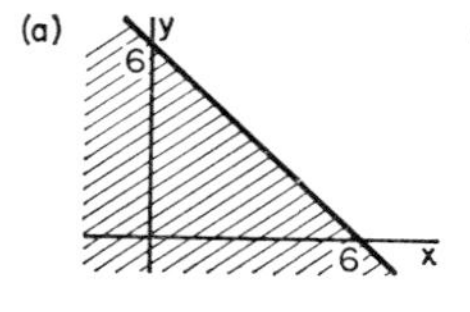

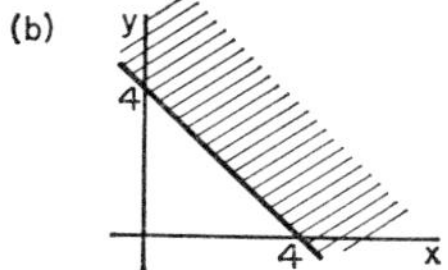

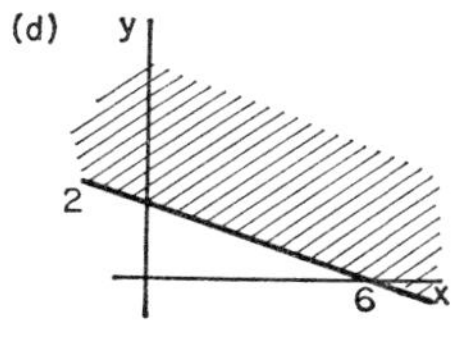

(c) y 6 2 x

5.

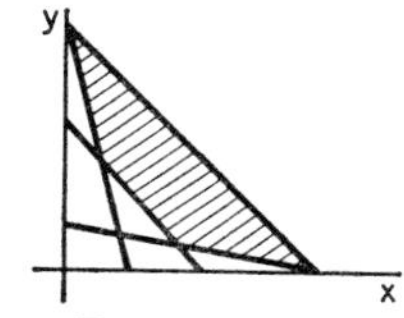

FIG. 2

6. (a) (0, 6) (b) (6, 0)
(c) (1, 3) (d) (3, 1)

7. (a) {(0, 6)} (b) $\{(1\frac{1}{2}, 1\frac{1}{2})\}$
(c) {(6, 0)} (d) {(1, 3)}
(e) {(3, 1)}

8.

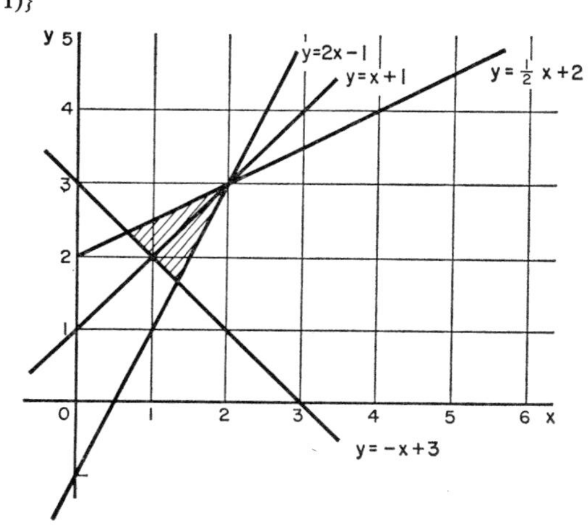

FIG. 3

9. (a) (2, 3)
(b) (2, 3)
(c) (1, 2)

10.

Region required is shaded in Fig. 3 above.

$$\{(x,y) \mid y > -x + 3\} \cap \{(x,y) \mid y > 2x - 1\} \cap \{(x,y) \mid y < \tfrac{1}{2}x + 2\}$$

Exercise 2(*d*)

1. (a) Function.
(b) Function.
(c) Not a function.
(d) Not a function.
(e) Function.

2. (a) Function.
(b) Function.
(c) Not a function.

3. Function.
4. Not a function.
5. Function.
6. Not a function.
7. Function.
8. Function.
9. Not a function.
10. Not a function.
11. Function.
12. Not a function.
13. Not a function.
14. Not a function.
15. Not a function.
16. Function.
17. Not a function.
18. Not a function.
19. Function.
20. Not a function.
21. Function.
22. Function.
23. Not a function.
24. Function.
25. Not a function.

Exercise 2(*e*)

In the following answers, R is the set of real numbers, R^+ is the set of positive real numbers. In each case the domain is given before the range.

1. 1:1; R; R.
2. Not 1:1; R; $\{4\}$.
3. 1:1; R; R.
4. Not 1:1; R; R^+.

5. 1:1; R/o: R/o.
6. Not 1:1; the set of integers; $\{0, 1\}$.
7. Not 1:1; R; R^+.
8. Not 1:1; R; all real numbers $\geqq 1$.
9. Not 1:1; $\{-2 \leqq x \leqq 2\}$; $\{0 \leqq y \leqq 2\}$.
10. Not 1:1; R; $\{0 \leqq y < 1\}$.
11. No.
12. (a) c (b) c (c) b (d) a
13. (a) a (b) b (c) a
14. (a) a (b) a (c) b (d) a (e) b
15. (a) c (b) c (c) c
16. f is one:one.
17. (a) b (b) d (c) a (d) e (e) c
18. (a) d (b) d (c) b (d) e (e) b
19. (a) e (b) e (c) d (d) c (e) d
20. (a) d (b) e (c) d (d) b (e) b
21. (a) Equal. (b) Equal. (c) Equal. (d) Unequal.
22. $f(x) = x$
23. $\{2, 3, 4, 5\}$; $\{\frac{1}{2}, \frac{1}{3}, \frac{1}{4}, \frac{1}{5}\}$.
24. (a) $\frac{1}{4}, \frac{1}{3}, -1$
(b) $-3, -2, 2$
$f \circ g(x) = g \circ f(x) = x$
25. $\text{Tan}^2 x$; $\tan(x^2)$.
26. (a) $f_4(x)$ (b) $f_3(x)$ (c) $f_1(x)$ (d) $f_4(x)$ (e) $f_1(x)$ (f) $f_1(x)$
27. (a) $f_2(x)$ (b) $f_2(x)$ (c) $f_1(x)$ (d) $f_1(x)$ (e) $f_5(x)$ (f) $f_6(x)$ (g) $f_1(x)$ (h) $f_1(x)$

CHAPTER 3

Exercise 3(*a*)

1. 3′ × 4′; 3′ × 5′; 4′ × 5′.
2. 8 cows, 12 sheep.
3. D_1 to C_1 20 tons; D_2 to C_2 15 tons; D_1 to C_3 10 tons.
4. $x = 100$, $y = 10$.
5. $x = 6$, $y = 2$.
6. $x = 20$, $y = 10$.
7. (5, 2) (6, 2) (6, 3) (7, 2); (6, 3) is the best solution.
8. (a) (4, 2) (b) (2, 4)
9. 41/–.
10. (5, 2) (7½, 1), or any ordered pair between, such that $2x + 5y = 20$.
11. D_1; 30 to C_1, 30 to C_2; D_2; 20 to C_1, 0 to C_2.

CHAPTER 4

Exercise 4(*a*)

1. F.
2. T.
3. F.
4. (a) Argument F but (b) conclusion true.
(c) R (d) R
(e) Isosceles trapezia, rectangles and squares.

5. (a) No. (b) Yes.
(c) Rhombuses, squares. (d) Squares.
(e) Squares.

6. F.
7. F.
8. T.
9. F.
10. F.
11. F.
12. T.
13. T.
14. F.
15. F.
16. F.
17. F.
18. F.
19. F.
20. T.

21. F. (Valid conclusion is $y < 0$.)
22. T.
23. No intelligent, red haired people are bad tempered.
24. Some old things are beautiful.
25. White cows do not wear bells.
26. Some people are neither sensible nor healthy.
27. Some complex numbers are natural numbers.

Exercise 4(*b*)

1. (a) I like tennis and I am a good cricketer.
(b) I either like tennis or I am a good cricketer (or both).
(c) I like tennis and I am not a good cricketer.
(d) Either I do not like tennis or I am a good cricketer (or both).
(e) I do not like tennis and I am not a good cricketer.
(f) I like tennis, therefore I am a good cricketer.

2. (a) History is bunk and religion is the opium of the people.
(b) Either history is bunk or religion is the opium of the people (or both).
(c) History is not bunk and religion is the opium of the people.
(d) Either history is bunk or religion is not the opium of the people (or both).
(e) Either history is not bunk or religion is the opium of the people (or both).
(f) History is bunk and religion is not the opium of the people.
(g) History is not bunk and religion is not the opium of the people.
(h) Either history is not bunk or religion is not the opium of the people (or both).
(i) "History is bunk" implies that religion is the opium of the people and vice-versa.

3. (a) I am not wearing a green hat.
(b) Either I am wearing a green hat or I am not wearing a green hat (or both). If we include (or both) the statement is nonsensical.
(c) Either I am wearing black shoes or I am not wearing black shoes (or both). Here again the inclusive disjunction makes nonsense of the statement. For this purpose $\vee$ should be used as an exclusive disjunction.
(d) Either I am wearing a green hat or I am not wearing black shoes.
(e) I am wearing a green hat and I am not wearing black shoes.
(f) I am wearing a green hat, therefore I am wearing black shoes.
(g) I am not wearing a green hat but I am wearing black shoes.
(h) Either I am wearing a green hat but not black shoes, or I am wearing black shoes but not a green hat (or both). Here again the inclusive disjunction can not be used.

4. (a) The two triangles are congruent and isosceles.
(b) The two triangles are isosceles and right angled.
(c) The triangles are congruent, isosceles and right angled.
(d) The triangles are either isosceles or similar (or both).
(e) The triangles are either both right angled or similar (or both).
(f) The triangles are either congruent or similar (or both).
(g) The triangles are congruent, therefore they are similar.
(h) The triangles are right angled, therefore they are both isosceles.
(i) The triangles are congruent and are both isosceles, hence they are similar.
(h) This is false.

5. (a) T (b) T (c) F (d) T
(e) T (f) F (g) T (h) T
(i) T (j) F (k) T (l) T
(m) T (n) F (p) F

6. (a) T (b) F (c) F (d) T
(e) F (f) F (g) F (h) T

Exercise 4(*c*)

1.

q	$\sim q$
T	F
F	T

2.

p	q	$\sim q$	$p \wedge \sim q$
T	T	F	F
T	F	T	T
F	T	F	F
F	F	T	F

3.

p	q	$\sim p$	$\sim p \wedge q$
T	T	F	F
T	F	F	F
F	T	T	T
F	F	T	F

4.

p	q	$\sim p$	$\sim q$	$\sim p \wedge \sim q$
T	T	F	F	F
T	F	F	T	F
F	T	T	F	F
F	F	T	T	T

In the following, the various possible values of p and q are taken in the same order as above, but only the value of the whole statement is given (e.g. 4 above could be abbreviated to read FFFT).

5. TT.
6. TTFT.
7. TFTT.
8. TFFT.
9. FTTT.
10. TFFT.
11. Both have truth table TFTT.
12. Both have truth set TFTT.
14. $p \to q$.
16. The simplest is $p \wedge \sim p$.

Exercise 4(*d*)

1. a.
2. a.
3. $a + b$.
4. 1
5. $a + b$.
6. $a + b$.
7. a
8. a.
9. 1.
10. $a.b$.
11. a.
12. $a'b$.
13. 0.
14. 1.
15. $a + b$.
16. $a + b$.
17. a.
18. a.

Exercise 4(*e*)

1. White cows do not wear bells. (It is *untrue* to say that Farmer Jones' cows do not wear bells.)
2. Boys not wearing ties should wear a white shirt.
3. One who is either dark, or, handsome but not dark.

Exercise (4*f*)

1. $x.x$ or x.
2. $x + x$ or x.
3. $x + xy$ or x.
4. $(x + x)(x + y)(x + z)$ or x.
5. $(x' + y')(x' + y)(x + y)$ or $x'y$.
6. $(x' + y')(x + y')(x + y)$ or xy'.
7. $(x + yz) + (x + yz)(x + yz)$ or $x + yz$.

8. $(x + y)(x + z)(x + y)(x + z)$ or $x + yz$.
9. $(a + b + c + xy)(a + b + c + x' + y')$ or $a + b + c$.
10. $(a + b)(a + c)(a + b' + c')$ or a.
11. $(a + bc)(a + b' + c')(a + b + c)(a + b + c')$ or a.
12. $abc + abc' + ab'c + a'bc$.

a	b	c	$abc + abc' + ab'c + a'bc$
1	1	1	1
1	1	0	1
1	0	1	1
0	1	1	1
1	0	0	0
0	1	0	0
0	0	1	0
0	0	0	0

Current only flows when 2 or 3 out of the three switches are closed. Thus if this current is used to illuminate a bulb, the light indicates a majority vote of a committee of three.

13. $abc + a'b'c + a'bc' + ab'c'$.
Closure table gives 10001110. Any successive change in a or b or c changes the current from off to on or vice-versa. Thus the circuit could be used to give independent control of a light from any one of three switches, i.e. a "three landing" switch arrangement.

14. $(p + q + r + pqr)(p + q + r + p' + q' + r')(p + q + r')(p + q')$ or p.
15. $\{[(p + q)r + pq]r + pr\} s + prs$ or $rs(p + q)$.
16. $[(x + y)z + xz] x + yz$ or $z(x + y)$.
17. $pq(a + b + p' + q') + (a + b)rs$ or $(a + b)(pq + rs)$.
18. $(x + y + z) + pq + x'y' + (p' + q' + z')$ or 1, i.e. current always flows no matter what the position of the switches.
19. See question 13.
20. See question 12.
21. $p(r + q)$.
22. $bcde + \Sigma b'cde + a\Sigma b'c'de$ or

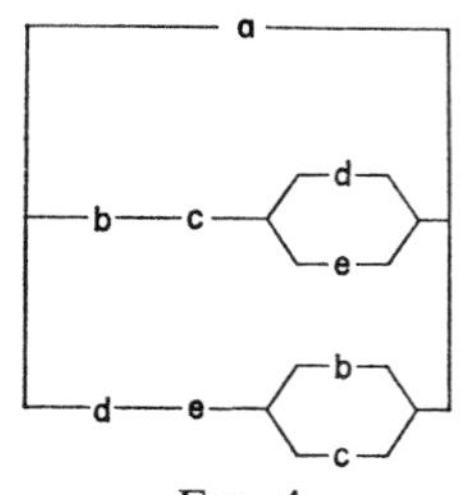

FIG. 4

23. $ac(b + d + e)$.

24. $abcd + abc + abd + acd + bcd + ab + ad + bc + cd$. This simplifies to $(a + c)(b + d)$ and the required fuse circuit is

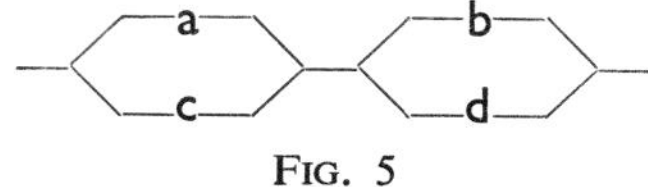

FIG. 5

CHAPTER 5

Exercise 5(a)

1. (a) 2 (b) 3 (c) 5 (d) 13
(e) 10 (f) 26 (g) 21 (h) 30
(i) 61 (j) 43 (k) 77 (l) 405
(m) 427 (n) 63 (o) 65 (p) 458
(q) 3634 (r) 5461

2. (a) 11 (b) 111 (c) 1011 (d) 11101
(e) 100111 (f) 111000 (g) 111111 (h) 1111110
(i) 10000100 (j) 1101111 (k) 11111111 (l) 1000000001
(m) 1001100000 (n) 1100000000 (o) 1100001001 (p) 10000000000
(q) 10011100010000 (r) 11110100001001000000

Exercise 5(b)

2. See Exercise 5(a) Question 2.

Exercise 5(c)

1. 1000; $5 + 3 = 8$.
2. 1100; $7 + 5 = 12$.
3. 10110; $13 + 9 = 22$.
4. 11001; $10 + 15 = 25$.
5. 11111; $21 + 10 = 31$.
6. 110010; $27 + 23 = 50$.
7. 1100001010; $429 + 349 = 778$.
8. 11011110100010; $7531 + 6711 = 14242$.
9. 100110; $13 + 11 + 14 = 38$.
10. 100000; $9 + 13 + 10 = 32$.
11. 110000; $11 + 15 + 13 + 9 = 48$.
12. 1100000; $29 + 23 + 17 + 27 = 96$.
13. 110010100; $111 + 99 + 79 + 115 = 404$.
14. 1111000; $63 + 31 + 15 + 7 + 3 + 1 = 120$.

15. 11110; $15 + 15 = 30$.
16. 111100; $15 + 15 + 15 + 15 = 60$.
17. 10; $5 - 3 = 2$.
18. 100; $9 - 5 = 4$.
19. 111; $16 - 9 = 7$.
20. 1011; $21 - 10 = 11$.
21. 1001; $16 - 7 = 9$.
22. 110000; $89 - 41 = 48$.
23. 100111; $102 - 63 = 39$.
24. 11001101100; $11061 - 9417 = 1644$.
25. 0; $30 - 15 - 15 = 0$.
26. 0; $15 - 5 - 5 - 5 = 0$.
27. 0; $33 - 11 - 11 - 11 = 0$.
28. 100000001; $512 - 255 = 257$.

Exercise 5(*d*)

1. 1100; $6 \times 2 = 12$.
2. 100100; $9 \times 4 = 36$.
3. 1111000; $15 \times 8 = 120$.
4. 10101; $7 \times 3 = 21$.
5. 1101110; $11 \times 10 = 110$.
6. 110110010; $31 \times 14 = 434$.
7. 11110100001; $63 \times 31 = 1953$.
8. 10000001011; $45 \times 23 = 1035$.
9. 110111110010; $85 \times 42 = 3570$.
10. 100010011100001; $229 \times 77 = 17633$.
11. 11; $6 \div 2 = 3$.
12. 10; $8 \div 4 = 2$.
13. 11; $24 \div 8 = 3$.
14. 101; $15 \div 3 = 5$.
15. 111; $63 \div 9 = 7$.
16. 1011; $121 \div 11 = 11$.
17. 110; $132 \div 22 = 6$.
18. 100; $140 \div 35 = 4$.
19. 10001; $289 \div 17 = 17$.
20. 1010; $1000 \div 100 = 10$.

Exercise 5(*e*)

1. (a) $1\frac{1}{2}$ (b) $3\frac{1}{4}$ (c) $5\frac{5}{8}$ (d) $\frac{5}{16}$
(e) $2\frac{13}{16}$ (f) $6\frac{11}{16}$ (g) $10\frac{27}{32}$ (h) $1\frac{3}{32}$
(i) $23\frac{7}{32}$ (j) $7\frac{7}{8}$ (k) $1\frac{31}{32}$ (l) $26\frac{27}{64}$

2. (a) 1·1 (b) 10.01 (c) 11·001 (d) 101·011
(e) 111·101 (f) 1011·1011 (g) 1110·11111 (h) 10001·000011
(i) 1111111·1111111

3. (a) 1·01 (b) 11·11 (c) 0·1 (d) 110·111
(e) 111·101 (f) 11·0001 (g) 111·1001 (h) 0·00101
(i) 0.010001

4. (a) ·00011 (b) ·00110 (c) ·01001 (d) ·1
(e) ·11101 (f) ·00100 (g) 10·11100 (h) 11.00100
(i) 1010·00000

5. (a) ·11 (b) 11·011 (c) 10·0111 (d) 1·10111
(e) ·10101 (f) 11·1100001 (g) 111·00001
(h) 10010·1010001 (i) 11 (j) 10110
(k) 101·1 (l) 1010·1 (m) 1001.1 (n) 10000
(o) 1101 (p) 1100·11

Exercise 5(*f*)

1. (a) 100 (b) 110 (c) 122 (d) 1220
(e) 11010

2. (a) 110 (b) 444 (c) 10224 (d) 10300
(e) 13000

3. (a) 30 (b) 104 (c) 240 (d) 1054
(e) 10010

4. (a) 70 (b) 77 (c) 115 (d) 144
(e) 1101

5. (a) 122 (b) t 8 (c) 561 (d) 133
(e) 615 (f) 552 (g) 132 (h) 1423
(i) 101

6. (a) 1000 (b) 41 (c) 3533

8. + ∋; 工 王 X̅ ; 189; 164; + 工 ∋

9. 388; 44; 442; 3.

CHAPTER 6

Exercise 6(*a*)

1. *B*, *C*, *D*, *E*, *H*, *I*, *K*, *O*, *X*.
2. *A*, *H*, *I*, *M*, *O*, *T*, *V*, *W*, *X*, *Y*.
3. *H*, *I*, *N*, *O*, *S*, *X*, *Z*.
4. *H*, *I*, *O*, *X*.

5. (a) ɿ (b) ɭ (c) ſ (d) J
(e) J (f) J (g) J (h) ſ
(i) ſ (j) ɭ (k) ɭ (l) ɿ
(m) ɿ (n) ɿ (o) ɭ (p) ſ
(q) J

$\times$	I	p	q	r
I	I	p	q	r
p	p	I	r	q
q	q	r	I	p
r	r	q	p	I

Exercise 6(*b*)

$\times$	1	ω	ω^2
1	1	ω	ω^2
ω	ω	ω^2	1
ω^2	ω^2	1	ω

2.

2nd operation \ 1st operation	I	ω	ω^2	p	q	r
I	I	ω	ω^2	p	q	r
ω	ω	ω^2	I	r	p	q
ω^2	ω^2	I	ω	q	r	p
p	p	q	r	I	ω	ω^2
q	q	r	p	ω^2	I	ω
r	r	p	q	ω	ω^2	I

3.

2nd operation \ 1st operation	1	ω	ω^2	ω^3	p	q	r	s
1	1	ω	ω^2	ω^3	p	q	r	s
ω	ω	ω^2	ω^3	1	r	s	q	p
ω^2	ω^2	ω^3	1	ω	q	p	s	r
ω^3	ω^3	1	ω	ω^2	s	r	p	q
p	p	s	q	r	1	ω^2	ω^3	ω
q	q	r	p	s	ω^2	1	ω	ω^3
r	r	p	s	q	ω	ω^3	1	ω^2
s	s	q	r	p	ω^3	ω	ω^2	1

Exercise 6(*c*)

1. Group.
2. Group.
3. Group.
4. Not a group.
5. Not a group.
6. Group.
7. Not a group.
8. Not a group.

9. Group.
10. Not a group.
11. Not a group.
12. Group.
13. Not a group.
14. Group.
15. Group.
16. Not a group.
17. Group.

Exercise 6(*d*)

$\times$	f_1	f_2	f_3	f_4
f_1	f_1	f_2	f_3	f_4
f_2	f_2	f_1	f_4	f_3
f_3	f_3	f_4	f_1	f_2
f_4	f_4	f_3	f_2	f_1

The identity element is f_1.

Exercise 6(*e*)

1. $\{f_1, f_2\}$, $\{f_1, f_3\}$, $\{f_1, f_4\}$.
2. $\{1, \omega^2, p, q\}$ (This is the group of the rectangle.)
   Other subgroups are $\{1, \omega, \omega^2, \omega^3\}$; $\{1, p\}$; $\{1, q\}$; $\{1, r\}$; $\{1, s\}$; $\{1, \omega^2\}$.
3. $\{I, p\}$; $\{I, q\}$; $\{I, r\}$.
4. The $\pm$ rational numbers (including 0); the integers (including 0).
5. The $\pm$ rational numbers; the set $\{\ldots n^{-3}, n^{-2}, n^{-1}, n^0, n^1, n^2, n^3$. where n is any positive integer.
6. $\{1, -1\}$.

7.

+	0	1	2	3	4	5
0	0	1	2	3	4	5
1	1	2	3	4	5	0
2	2	3	4	5	0	1
3	3	4	5	0	1	2
4	4	5	0	1	2	3
5	5	0	1	2	3	4

Subgroups are $\{0, 2, 4\}$ and $\{0, 3\}$.

8.

$\times$	1	3	5	7
1	1	3	5	7
3	3	1	7	5
5	5	7	1	3
7	7	5	3	1

mod 8

The subgroups are $\{1, 3\}$; $\{1, 5\}$; $\{1, 7\}$.

Exercise 6(*f*)

4. The group of rotations $\{1, \omega, \omega^2 \ldots \omega^{n-1}\}$, where $\omega^n = 1$ and n is any positive integer, is isomorphic with the group of integers $\{0, 1, 2, \ldots n-1\}$ modulo n under addition.

5. The subgroup of order 4 is $\{0,2,4,6\}$ and this itself has a subgroup of order 2, $\{0,4\}$.

6. $\{1,3\}$, $\{f_1,f_2\}$; $\{1, 5\}$, $\{f_1, f_3\}$; $\{1, 7\}$, $\{f_1, f_4\}$. N.B. Each group, apart from the group with only one element, has two improper subgroups, itself and the identity. These cases have not been included here.

7. A, D, E, C, B, F, with F the hardest. The more symmetrical the cross sectional figure, the greater is the number of operations or rotations by which the block can be posted. The greater the choice of possible operations, the greater is the chance that the child will find one of them. The group of symmetries of a figure with many axes of symmetry is larger than that of one with few. The group of symmetries of A is an infinite cyclic group, and posting through A is by far the easiest operation. The others, being regular figures, all have cyclic groups. Thus, considering only rotations of each figure in its plane:—
for D we have $\{1, \omega, \omega^2, \omega^3, \omega^4, \omega^5\}$ where $\omega^6 = 1$,
for E we have $\{1, \omega, \omega^2, \omega^3, \omega^4\}$ where $\omega^5 = 1$,
for C we have $\{1, \omega, \omega^2, \omega^3\}$ where $\omega^4 = 1$,
for B we have $\{1, \omega, \omega^2\}$ where $\omega^3 = 1$.
If F is isosceles then $\{1,\omega\}$ where $\omega^2 = 1$, but if F is scalene it has the trivial group $\{1\}$. No two of these are isomorphic, but, for example, D's subgroup $\{1, \omega^2, \omega^4\}$ is isomorphic with B's group $\{1,\omega,\omega^2\}$, C's subgroup $\{1,\omega^2\}$ is isomorphic with F's group $\{1,\omega\}$.

CHAPTER 7

Exercise 7(*a*)

1. $\begin{pmatrix} -1 & 0 \\ 0 & 1 \end{pmatrix}$

2. $\begin{pmatrix} -1 & 0 \\ 0 & -1 \end{pmatrix}$

3. $\begin{pmatrix} 1 & 0 \\ 0 & -1 \end{pmatrix}$

4. $\begin{pmatrix} -1 & 0 \\ 0 & 1 \end{pmatrix}$

5. $\begin{pmatrix} 1 & 0 \\ 0 & 1 \end{pmatrix}$

Exercise 7(*b*)

1. $\begin{pmatrix} 10 & 7 \\ 12 & 9 \end{pmatrix}$; $\begin{pmatrix} 2 & 7 \\ 4 & 17 \end{pmatrix}$

2. $\begin{pmatrix} 4 & 3 \\ 7 & 4 \end{pmatrix}$; $\begin{pmatrix} 8 & 1 \\ 5 & 0 \end{pmatrix}$

3. $\begin{pmatrix} 7 & -6 \\ -19 & 12 \end{pmatrix}$; $\begin{pmatrix} 5 & -10 \\ -10 & 14 \end{pmatrix}$

4. $\begin{pmatrix}1 & 0\\0 & 1\end{pmatrix}$ P maps any point onto its reflection in the X axis. Applied twice, any point is mapped back onto itself.

$P^{10} = \begin{pmatrix}1 & 0\\0 & 1\end{pmatrix}$ $\quad P^{11} = \begin{pmatrix}1 & 0\\0 & -1\end{pmatrix}$

5. (a) $\begin{pmatrix}1 & 0\\0 & 1\end{pmatrix}$ (b) $\begin{pmatrix}1 & 0\\0 & -1\end{pmatrix}$

(c) $\begin{pmatrix}-1 & 0\\0 & 1\end{pmatrix}$ (d) $\begin{pmatrix}-1 & 0\\0 & -1\end{pmatrix}$

(e) $\begin{pmatrix}a & b\\c & d\end{pmatrix}$ (f) $\begin{pmatrix}0 & 0\\0 & 0\end{pmatrix}$

6. $\begin{pmatrix}0 & 0\\0 & 0\end{pmatrix}$, $\begin{pmatrix}-1 & -1\\1 & 1\end{pmatrix}$ $\quad A.B \neq B.A$

Further, $A.B$ is "zero" (the null matrix), but neither A nor B are "zero" (null matrices). In both cases

$$|A|.|B| = |A.B|$$
$$\text{and } |B|.|A| = |B.A|$$

7. (a) $B = \begin{pmatrix}-1 & 0\\0 & -1\end{pmatrix}$ (b) $C = \begin{pmatrix}0 & 1\\-1 & 0\end{pmatrix}$

(c) $I = \begin{pmatrix}1 & 0\\0 & 1\end{pmatrix}$

8. (a) (1, 1) (2, 2); PQ is unaltered.
(b) (1, −1) (2, −2); PQ is mapped onto its reflection in OX.
(c) (−1, 1) (−2, 2); PQ is mapped onto its reflection in OY.
(d) $(0, \sqrt{2})$ $(0, 2\sqrt{2})$; PQ is rotated anti-clockwise about 0 through 45°.
(e) (0, 2) (0, 4); OP and OQ are each rotated anti-clockwise about 0 through 45° and increased in length $\sqrt{2}$ times.
(f) (2, 2) (4, 4); OP and OQ and hence PQ are doubled in length without rotation, or P is translated to (2,2) and Q is translated to (4,4).
(g) (3, 3) (6, 6); OP, OQ and hence PQ are trebled in length without rotation.

9. $\begin{pmatrix}x_3\\y_3\end{pmatrix} = \begin{pmatrix}7 & -1\\1 & 7\end{pmatrix} . \begin{pmatrix}x_0\\y_0\end{pmatrix}$ i.e. $\begin{matrix}x_3 = 7x_0 - y_0\\ y_3 = x_0 + 7y_0\end{matrix}$

10. (a) $\begin{pmatrix}2 & 0\\0 & 2\end{pmatrix}$ (b) $\begin{pmatrix}3 & 0\\0 & 3\end{pmatrix}$

(c) $\begin{pmatrix}1 & 0\\0 & -1\end{pmatrix}$ (d) $\begin{pmatrix}-1 & 0\\0 & 1\end{pmatrix}$

(e) $\begin{pmatrix}-1 & 0\\0 & -1\end{pmatrix}$ (f) $\begin{pmatrix}0 & -1\\1 & 0\end{pmatrix}$

(g) $\begin{pmatrix}\frac{1}{\sqrt{2}} & -\frac{1}{\sqrt{2}}\\ \frac{1}{\sqrt{2}} & \frac{1}{\sqrt{2}}\end{pmatrix}$ (h) $\begin{pmatrix}0 & -2\\2 & 0\end{pmatrix}$

11. The square is transformed into a rhombus, vertices (0,0); $(1,\frac{1}{2})$; $(1\frac{1}{2},1\frac{1}{2})$; $(\frac{1}{2},1)$.

12. The square is transformed into a line segment $\overrightarrow{OQ}$ where Q is the point (2, 2).

Exercise 7(*c*)

1. $(5,6)\begin{pmatrix}3 & 7\\2 & 9\end{pmatrix} = (27,89)$ i e. 27 units of vitamin A, 89 units of vitamin B.

$(5,6)\begin{pmatrix}3\\3\frac{1}{2}\end{pmatrix} = (36)$ i.e. the cost is 36/–.

2. (i) $\begin{pmatrix}20 & 30\\40 & 10\end{pmatrix}\cdot\begin{pmatrix}5\\7\end{pmatrix} = \begin{pmatrix}310\\270\end{pmatrix}$ i.e. 310 saloons, 270 vans.

(ii) $(500,400)\,.\begin{pmatrix}20 & 30\\40 & 10\end{pmatrix}\cdot\begin{pmatrix}5\\7\end{pmatrix} = (500,400)\,.\begin{pmatrix}310\\270\end{pmatrix} = (263{,}000)$

i.e. the total weekly value = £263,000.

3. (i) $(9,6)\begin{pmatrix}1{,}600 & 2{,}000\\1{,}500 & 1{,}800\end{pmatrix} = (23{,}400,\ 28{,}800)$

i.e. 23,400 units of materials and 28,800 hours of labour.

(ii) $£\begin{pmatrix}1{,}600 & 2{,}000\\1{,}500 & 1{,}800\end{pmatrix}\cdot\begin{pmatrix}1\\\frac{1}{2}\end{pmatrix} = £\begin{pmatrix}2{,}600\\2{,}400\end{pmatrix}$

i.e. a house costs £2,600, a bungalow £2,400.

(iii) $(9,6)\begin{pmatrix}1{,}600 & 2{,}000\\1{,}500 & 1{,}800\end{pmatrix}\cdot\begin{pmatrix}1\\\frac{1}{2}\end{pmatrix}$ pounds $= £\,(9,6)\begin{pmatrix}2{,}600\\2{,}400\end{pmatrix}$

= £37,800.

4. (i) $\begin{pmatrix}1 & 3\\2 & 1\frac{1}{2}\end{pmatrix}\cdot\begin{pmatrix}50\\200\end{pmatrix} = \begin{pmatrix}650\\400\end{pmatrix}$

i.e. £650, £400 per day respectively for each TV1, TV2 station.

(ii) $(2,3)\begin{pmatrix}1 & 3\\2 & 1\frac{1}{2}\end{pmatrix} = (8,10\frac{1}{2})$

i.e. 8 hours documentary and $10\frac{1}{2}$ hours variety.

(iii) $£\,(2,3)\begin{pmatrix}1 & 3\\2 & 1\frac{1}{2}\end{pmatrix}\cdot\begin{pmatrix}50\\200\end{pmatrix} = £\,(2,3)\begin{pmatrix}650\\400\end{pmatrix} = £2{,}500$

5. $£\,(3{,}000,\ 2{,}000,\ 1{,}000)\,.\begin{pmatrix}40 & 100 & 50\\80 & 150 & 80\\100 & 250 & 100\end{pmatrix}\cdot\begin{pmatrix}2\\\frac{1}{2}\\1\end{pmatrix}$

= £1,595,000.

Exercise 7(*d*)

1. $2x^2 - 5xy + 2y^2 = 0$ i.e. $x = y/2$ or $x = 2y$.

2. $a = 1$ or 3.

3. (i) $x^2 + 4y^2 = 4$ ellipse (ii) $x^2 - y^2 = 1$

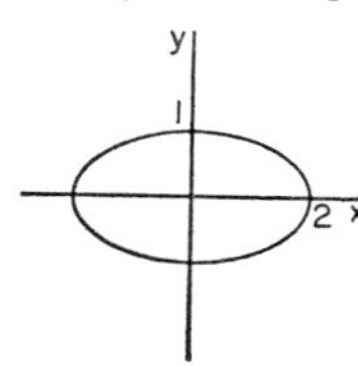

FIG. 6

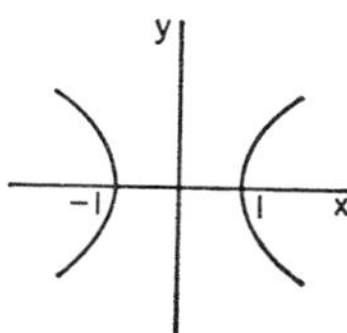

FIG. 7

4. (a)

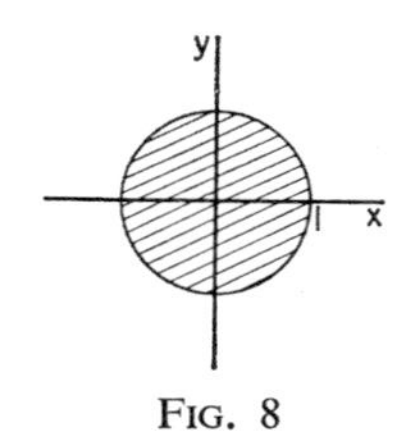

FIG. 8

(b)

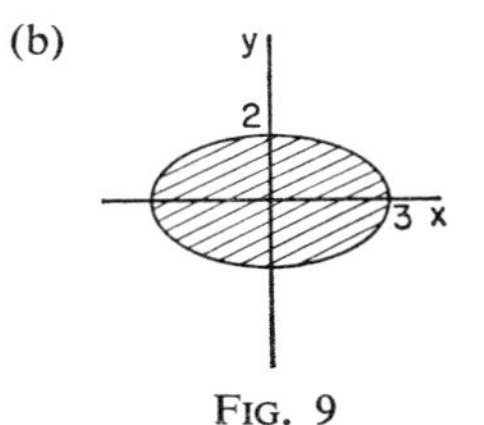

FIG. 9

5. $X^2 + \dfrac{Y^2}{k^2} = 1$

(a) $k = \frac{1}{2}$ (b) $k = \sqrt{3}/2$

Exercise 7(*e*)

1. $x = 5, y = 1, z = 3$.
2. $x = 2, y = 1, a = 4, b = 3$.
3. $x = 1, y = 2, z = 3, w = 4$.
4. $x = 4, y = 3$.
5. $x = 4$ or -1, $y = 2$ or -1, $z = 2$ (twice), $w = 1$ or 0.

Exercise 7(*f*)

1. (a) $\begin{pmatrix} 5 & 1 \\ 4 & 7 \end{pmatrix}$ (b) $\begin{pmatrix} 4 & 2 \\ 5 & 5 \end{pmatrix}$ (c) $\begin{pmatrix} 7 & -3 \\ 1 & 8 \end{pmatrix}$

(d) $\begin{pmatrix} 2 & 6 \\ 8 & 4 \end{pmatrix}$ (e) $\begin{pmatrix} 12 & -6 \\ 0 & 15 \end{pmatrix}$ (f) $\begin{pmatrix} 12 & -4 \\ 4 & 12 \end{pmatrix}$

(g) $\begin{pmatrix} -3 & 5 \\ 4 & -3 \end{pmatrix}$ (h) $\begin{pmatrix} -2 & 0 \\ -2 & -1 \end{pmatrix}$ (i) $\begin{pmatrix} 2 & 4 \\ 4 & 7 \end{pmatrix}$

2. (a) $\begin{pmatrix} 5 & -2 \\ -3 & -4 \end{pmatrix}$ (b) $\begin{pmatrix} 2 & 3 \\ 4 & 4 \end{pmatrix}$ (c) $\begin{pmatrix} 3 & 2 & -1 \\ -2 & -1 & -2 \\ 1 & -1 & 5 \end{pmatrix}$

(d) $\begin{pmatrix} 2 \\ 8 \\ -1 \end{pmatrix}$ (e) $(3, 1, -13)$

3. (a) $\begin{pmatrix} 4 \\ 4 \end{pmatrix}$ (b) $\begin{pmatrix} -2 \\ 2 \end{pmatrix}$ (c) $\begin{pmatrix} 2 \\ 2 \end{pmatrix}$

(d) $\begin{pmatrix} 5 \\ 7 \end{pmatrix}$ (e) $\begin{pmatrix} 1\frac{2}{3} \\ 2\frac{1}{3} \end{pmatrix}$ (f) $\begin{pmatrix} 2\frac{1}{3} \\ 1\frac{2}{3} \end{pmatrix}$

(g) $\begin{pmatrix} 0 \\ 8 \end{pmatrix}$ (h) $\begin{pmatrix} -8 \\ 0 \end{pmatrix}$ (i) $\begin{pmatrix} 2\frac{1}{5} \\ 1\frac{4}{5} \end{pmatrix}$

If X is the *point* (1,3), Y is the *point* (3,1), then:

(c) is the midpoint of XY,

(e) is the point of internal trisection of XY nearer to X,

(f) is the point of internal trisection of XY nearer to Y,

(i) is the point which divides $\overrightarrow{XY}$ internally in the ratio 3:2 (i.e. $\frac{3}{5}$ of the way along XY from X).

5. (d) In general, if $\overrightarrow{OS} = \dfrac{lP + mQ}{l + m}$ then S divides $\overrightarrow{PQ}$ in the ratio $m{:}l$.

Exercise 7(*g*)

1. (a) $\begin{pmatrix} 2 & -3 \\ -1 & 2 \end{pmatrix}$ (b) $\begin{pmatrix} 1 & -1 \\ -2 & 3 \end{pmatrix}$ (c) $\begin{pmatrix} -1 & 0 \\ 0 & -1 \end{pmatrix}$

d) $\begin{pmatrix} 1 & 0 \\ 0 & -1 \end{pmatrix}$ (e) $\begin{pmatrix} -1 & 0 \\ 0 & 1 \end{pmatrix}$ (f) $\begin{pmatrix} 1 & 0 \\ 0 & 1 \end{pmatrix}$

(g) $-\frac{1}{14}\begin{pmatrix} -1 & -3 \\ -4 & 2 \end{pmatrix}$ (h) $\frac{1}{10}\begin{pmatrix} 1 & 2 \\ -3 & 4 \end{pmatrix}$

(c), (d), (e) and (f) are each their own inverses. (See the group of movements of the rectangle.)

2. (a), (b) and (c) do not possess inverses. Matrices $\begin{pmatrix} a & b \\ c & d \end{pmatrix}$, where $ad - bc = 0$ $\left(\text{or} \begin{vmatrix} a & b \\ c & d \end{vmatrix} = 0\right)$, do not possess inverses. They are called *singular* matrices.

Take any point (x, y) in the plane. This is mapped by the matrix $\begin{pmatrix} a & b \\ c & d \end{pmatrix}$ into the point (X, Y) where

$$\begin{pmatrix} X \\ Y \end{pmatrix} = \begin{pmatrix} a & b \\ c & d \end{pmatrix}\begin{pmatrix} x \\ y \end{pmatrix} \quad \text{or} \quad \begin{matrix} X = ax + by \\ Y = cx + dy \end{matrix}$$

Now $cX - aY = (bc - ad)y = 0$ if $ad - bc = 0$ and y is finite, i.e. all points in the plane are mapped onto a straight line $Y = \frac{c}{a}X$ by the singular matrix $\begin{pmatrix} a & b \\ c & d \end{pmatrix}$.

Exercise 7(*h*)

1. (a) $x = 1, y = 2$ (b) $x = 2, y = 3$ (c) $x = -1, y = 3$
(d) $x = 1, y = 2$ (e) $x = 1, y = -1$ (f) $x = 2, y = -1$
(g) $x = 3, y = 1$ (h) $x = p + q, y = p - q$

2. $\frac{1}{10}\begin{pmatrix} 1 & 2 \\ -3 & 4 \end{pmatrix}$ (a) $x = \frac{1}{10}(p + 2q), y = \frac{1}{10}(-3p + 4q)$
(b) $x = 4, y = 2$ (c) $x = 1, y = 2$ (d) $x = 1, y = -\frac{1}{2}$

3. (a) $\begin{pmatrix} 2 & -3 \\ -3 & 5 \end{pmatrix}$ (b) $\begin{pmatrix} 1 & -1 \\ -1 & 2 \end{pmatrix}$ (c) $\begin{pmatrix} 13 & 8 \\ 8 & 5 \end{pmatrix}$
(d) $\begin{pmatrix} 13 & 8 \\ 8 & 5 \end{pmatrix}$ (e) $\begin{pmatrix} 5 & -8 \\ -8 & 13 \end{pmatrix}$ (f) $\begin{pmatrix} 5 & -8 \\ -8 & 13 \end{pmatrix}$
(g) $\begin{pmatrix} 5 & -8 \\ -8 & 13 \end{pmatrix}$ (h) $\begin{pmatrix} 1 & 0 \\ 0 & 1 \end{pmatrix}$

$A.B = B.A$ (These two matrices are commutative but in general matrix multiplication is non-commutative.)

4. (a) $\begin{pmatrix} 2 & -3 \\ -1 & 2 \end{pmatrix}$ (b) $\begin{pmatrix} 2 & -1 \\ -1 & 1 \end{pmatrix}$ (c) $\begin{pmatrix} 5 & 8 \\ 3 & 5 \end{pmatrix}$
(d) $\begin{pmatrix} 3 & 5 \\ 4 & 7 \end{pmatrix}$ (e) $\begin{pmatrix} 5 & -8 \\ -3 & 5 \end{pmatrix}$ (f) $\begin{pmatrix} 5 & -8 \\ -3 & 5 \end{pmatrix}$
(g) $\begin{pmatrix} 7 & -5 \\ -4 & 3 \end{pmatrix}$ (h) $\begin{pmatrix} 7 & -5 \\ -4 & 3 \end{pmatrix}$

$AB \neq BA$, but $(AB)^{-1} = B^{-1} . A^{-1}$ and $(BA)^{-1} = A^{-1} . B^{-1}$.

5. (a) $\begin{pmatrix}8, & 7\\ 7, & 8\end{pmatrix}$ (b) $\frac{1}{3}\begin{pmatrix}2 & -1\\ -1 & 2\end{pmatrix}$ (c) $\frac{1}{5}\begin{pmatrix}3 & -2\\ -2 & 3\end{pmatrix}$

(d) $\frac{1}{15}\begin{pmatrix}8 & -7\\ -7 & 8\end{pmatrix}$

The magic number of A^{-1} is $\frac{1}{3}$.
The magic number of B is 5 and the magic number of B^{-1} is $\frac{1}{5}$.
The magic number of $A.B$ is 15 and the magic number of $(AB)^{-1}$ is $\frac{1}{15}$.

6. (a) $\begin{pmatrix}2 & 1\\ 3 & 4\end{pmatrix}$ (b) $\begin{pmatrix}1 & 2\\ 4 & 3\end{pmatrix}$ (c) $\begin{pmatrix}8 & 17\\ 9 & 16\end{pmatrix}$

(d) $\begin{pmatrix}8 & 9\\ 17 & 16\end{pmatrix}$ (e) $\begin{pmatrix}6 & 7\\ 19 & 18\end{pmatrix}$ (f) $\begin{pmatrix}8 & 9\\ 17 & 16\end{pmatrix}$

$\widetilde{P.Q} = \widetilde{Q}.\widetilde{P}$. Also $\widetilde{Q.P} = \widetilde{P}.\widetilde{Q}$.

7. $25X^2 - 10XY + 5Y^2 = 1$
$X^2 + Y^2 = 1$
The equation is unaltered because the matrix merely rotates every radius of the circle through 45° about the origin. You may like to show that in general the matrix $\begin{pmatrix}\cos\theta & -\sin\theta\\ \sin\theta & \cos\theta\end{pmatrix}$, which rotates each radius through any angle θ, leaves the equation of the circle unaltered.

8. The decoder obtains the incorrect message NEON.
(a) VYNL (b) YULP (c) BWRY
(d) GYDV (e) BOMB (f) ATOM
(g) JETS (h) SHIP

CHAPTER 8

Exercise 8(*a*)

1. (a) $(3, 6\frac{1}{2})$ (b) $(6, 7\frac{1}{2})$ (c) $(9, 6)$ (d) $(6, 1)$
(e) $(2, 3\frac{1}{2})$ (f) $(5, 2)$ (g) $(3, \frac{1}{2})$ (h) $(-3, 4)$
(i) $(-5, \frac{1}{2})$ (j) $(-1, -3)$ (k) $(-3, -1)$ (l) $(0, -6\frac{1}{2})$

2. (a) $\overrightarrow{DB}$ (b) $\overrightarrow{IF}$ (c) $\overrightarrow{EH}$

3. (a) $\overrightarrow{AB}$ or $\overrightarrow{FG}$ (b) $\overrightarrow{CB}$ (c) $\overrightarrow{BG}$

(d) $\overrightarrow{EC}$, $\overrightarrow{CD}$ or $\overrightarrow{ED}$ (e) $\overrightarrow{HI}$, $\overrightarrow{FC}$ or $\overrightarrow{EO}$

(f) $\overrightarrow{OA}$, $\overrightarrow{OC}$, $\overrightarrow{AB}$ or $\overrightarrow{FG}$ (g) $\overrightarrow{HJ}$

4. (a) $\overrightarrow{OF}$ (b) $\overrightarrow{FH}$ (c) $\overrightarrow{CH}$ (d) $\overrightarrow{BJ}$
(e) $\overrightarrow{BF}$ (f) $\overrightarrow{OB}$ (g) $\overrightarrow{OC}$ (h) $\overrightarrow{OC}$
(i) $\overrightarrow{BI}$ (j) $\overrightarrow{IB}$ (k) null vector
(l) $\overrightarrow{GC}$ (m) $\overrightarrow{OB}$ (n) $\overrightarrow{OE}$
(o) null vector (p) null vector (q) null vector

5. (a) (2, 2) (b) $(5, \frac{1}{2})$ (c) (−1, 2) (d) (2, 0)
(e) $(2, 1\frac{1}{2})$ (f) (6, −1) (g) (8/3, 8/3) (h) (0, 0)

6. (a) $\overrightarrow{AC}.\overrightarrow{OB} = (2, -2)\begin{pmatrix}4\\4\end{pmatrix} = 0$

(b) $\overrightarrow{OH}.\overrightarrow{BD} = (9, 6)\begin{pmatrix}2\\-3\end{pmatrix} = 0$

(c) $\overrightarrow{OE}.\overrightarrow{EC} = (0, 1)\begin{pmatrix}3\\0\end{pmatrix} = 0$

(d) $\overrightarrow{CD}.\overrightarrow{DG} = (3, 0)\begin{pmatrix}0\\6\frac{1}{2}\end{pmatrix} = 0$

Exercise 8(*b*)

1. (a) $\mathbf{p} + \mathbf{q}$ (b) $\mathbf{q} + \mathbf{r}$ (c) $\mathbf{p} + \mathbf{q} + \mathbf{r}$
(d) $\mathbf{q} + \mathbf{r} + \mathbf{s}$ (e) $\mathbf{p} + \mathbf{q} + \mathbf{r} + \mathbf{s}$
(f) $\frac{1}{2}(\mathbf{p} - \mathbf{q} - \mathbf{r} - \mathbf{s}$ (g) $\frac{1}{2}(\mathbf{p} + \mathbf{q} - \mathbf{r} - \mathbf{s})$
(h) $\frac{1}{2}(\mathbf{p} + \mathbf{q} + \mathbf{r} - \mathbf{s})$

2. (a) $-\mathbf{a}$ (b) $-\mathbf{b}$ (c) $-\mathbf{c}$ (d) $\mathbf{a}$
(e) $\mathbf{b}$ (f) $\mathbf{c}$ (g) $\mathbf{a} + \mathbf{b}$ (h) $\mathbf{a} + \mathbf{b} + \mathbf{c}$
(i) $\mathbf{b} + \mathbf{c}$

3. (a) $2\mathbf{a} + \mathbf{b}$ (b) $2\mathbf{b}$ (c) $2\mathbf{b}$ (d) $2\mathbf{a}$
(e) $2\mathbf{a} + 2\mathbf{b}$
$p = 2, q = 1, k = 1.$
$\overrightarrow{AC} = 2\mathbf{a} + 2\mathbf{b}$, $\overrightarrow{BF} = 2\mathbf{a} + 2\mathbf{b}$.

4. $\overrightarrow{OR} = \frac{1}{2}(\mathbf{p} + \mathbf{q})$, $\overrightarrow{OS} = \frac{1}{3}(\mathbf{p} + 2\mathbf{q})$. $\overrightarrow{RS} = \frac{1}{6}(\mathbf{q} - \mathbf{p})$.

5. (i) $\frac{1}{2}(\mathbf{a} + \mathbf{c})$ (ii) $\frac{1}{2}(\mathbf{b} + \mathbf{d})$
$\overrightarrow{DA} = \mathbf{a} - \mathbf{d}$, $\overrightarrow{CB} = \mathbf{b} - \mathbf{c}$.
(a) $\frac{1}{2}(\mathbf{a} + \mathbf{b})$ (b) $\frac{1}{3}(\mathbf{a} + \mathbf{b} + \mathbf{c})$
(c) $\frac{1}{4}(\mathbf{a} + \mathbf{b} + \mathbf{c} + \mathbf{d})$

14. The identity element is the null vector. The inverse of $l\mathbf{a} + m\mathbf{b}$ is $-l\mathbf{a} - m\mathbf{b}$

Exercise 8(*c*)

1. 12·65 m.p.h. in a direction N 71° 34′ W.

2. S 63° 26′ E.

3. (a) $\frac{1}{12}$ mile; 5 minutes; 3·16 m.p.h.

(b) He should attempt to swim in a direction making 19° 28′ with $\overrightarrow{AB}$. His speed will be 2·83 m.p.h. and his time will be 5·3 minutes.

4. Acceleration is 2·83 ($2\sqrt{2}$) inches per second at 45° to $\overrightarrow{AO}$.

Velocity is 5 ins. per second at 36° 52′ to $\overrightarrow{AO}$.

5. 8·058 miles in a direction N 37° 15′ E.
The man takes 2 hr 15 min. and the crow takes 40·3 min., i.e. the crow takes 94·7 minutes less.

6. 14° 2′; 25° 52′.

7. Normal time is 2 hrs. Actual time is 2 hrs 6 mins. (Speeds along the sides are 232 and 161 m.p.h.)

Exercise 8(*e*)

1. (a) $\mathbf{i} + 3\mathbf{j}$ (b) $3\mathbf{i} + \mathbf{j}$ (c) $4\mathbf{i} + 4\mathbf{j}$ (d) $2\mathbf{i} - 2\mathbf{j}$
(e) $9\mathbf{i} + 6\mathbf{j}$ (f) $2\mathbf{i} - 3\mathbf{j}$ (g) $\mathbf{j}$ (h) $3\mathbf{i}$
(i) $3\mathbf{i}$ (j) $6\frac{1}{2}\mathbf{j}$

2. (a) $(3\frac{1}{2}, 2\frac{1}{2})$ (b) $(3, 3\frac{2}{3})$ (c) $(8, 3\frac{1}{12})$ (d) $(5\frac{1}{7}, 6)$
(e) $(5\frac{1}{2}, 5\frac{1}{4})$

3. (a) $\mathbf{r} = \lambda(\mathbf{i} + 3\mathbf{j}) + (1 - \lambda)(4\mathbf{i} + 4\mathbf{j})$
(b) $\mathbf{r} = \lambda(3\mathbf{i} + 6\frac{1}{2}\mathbf{j}) + (1 - \lambda)(4\mathbf{i} + 4\mathbf{j})$
(c) $\mathbf{r} = \lambda(6\mathbf{i} + 7\frac{1}{2}\mathbf{j}) + (1 - \lambda)(9\mathbf{i} + 3\frac{1}{2}\mathbf{j})$
(d) $\mathbf{r} = \lambda(9\mathbf{i} + 6\mathbf{j}) + (1 - \lambda)(9\mathbf{i} + 3\frac{1}{2}\mathbf{j})$
(e) $\mathbf{r} = \lambda(3\mathbf{i} + \mathbf{j}) + (1 - \lambda)(6\mathbf{i} + \mathbf{j})$

4. (a) 6 (b) $1\frac{1}{4}$ (c) -4

5. The scalar product in each case is zero.

6. (a) $\cos^{-1}\frac{3}{5}$ or 53° 8′. (b) $\cos^{-1}\dfrac{5}{\sqrt{26}}$ or 11° 18′.
(c) $\cos^{-1}\dfrac{-1}{\sqrt{2}}$ or 135°. (d) $\cos^{-1}\dfrac{41}{13\sqrt{10}}$ or approx. 4°.

7. (a) (8/3, 8/3) (b) (16/3, 14/3) (c) $(7, 4\frac{1}{6})$

8. (a) 1 (b) 1 (c) 1 (d) 0
(e) 0 (f) 0

9. The resultant force is of magnitude 13 units and makes angles $\cos^{-1}\frac{3}{13}$, $\cos^{-1}\frac{4}{13}$, $\cos^{-1}\frac{12}{13}$ with $\overrightarrow{OX}$, $\overrightarrow{OY}$, $\overrightarrow{OZ}$ respectively.

10. $\cos^{-1}\frac{1}{3}$ or 70° 32′.

CHAPTER 9

Exercise 9(*a*)

1. (a) Angles are 120°, 90°, 150°.
(b) Column heights are in the ratio 4:3:5.

2. (a) Angles are 120°, 48°, 72°, 24°, 96°.
(b) Column heights are in the ratio 5:2:3:1:4.

3. (a) Angles are 162°, 126°, 54°, 18°.
(b) Column heights are in the ratio 9:7:3:1.

4. (a) Angles are approximately 205°, 32°, 123°.
(b) Column heights are approximately in the ratio 6·4:1:3·9.

5. (a) Angles are 80°, 120°, 100°, 60°.
(b) Column heights are in the ratio 4:6:5:3.

Exercise 9(*b*)

1. (a) $3\frac{1}{2}$ lb (b) $3\frac{1}{2}$ lb (c) 3·625 lb
2. 35 yrs.
3. 3·96 hrs.
5. Modal range is 8–10 stones.
Median value falls in the range 8–10 stones.
Mean value is approximately 9 stones (8·98 st.).
6. Mean value is 4·2 (4·17).
7. Mode 13·29 gm.
Median 13·30 gm.
Mean 13·30 gm.
8. The 18·8 reading is clearly a wrong reading—it differs so much from the others. In practice we should take this measurement again, but failing this we ignore it. 20·8 is a reasonable mean.
9. Median value 16·45 cc.
Mean value 16·50 cc.
10. Mean mark 47.
11. Mean mark 54;
Median mark 56.
12. Yes. Mean of first set 6. Mean of seco1.
13.

No. of heads	0	1	2	3	4
Frequency	1	4	6	4	1
No. of heads	0	1	2	3	4
Frequency	1	5	10	10	5

Exercise 9(*c*)

1. $\frac{1}{2}(4-3) = \frac{1}{2}$ lb.
3. $\frac{1}{2}(4\cdot3-4\cdot0) = \cdot15$
4. $\frac{1}{2}(21-20\cdot6) = \cdot2$
5. The range is 80 marks (7–87).
The semi-interquartile range is $\frac{1}{2}(57-38)$ or $9\frac{1}{2}$ marks.

6. The range is 77 marks (12–89).
The semi-interquartile range is $\frac{1}{2}(65-44)$ or $10\frac{1}{2}$ marks.

7. (i) Approximately 2.
(ii) Approximately $5\frac{1}{2}$.

8. (i) Mean deviation 2.
(ii) Mean deviation 5·25.

9. ·23

10. ·03

11. ·14

12. (i) Standard deviation 3·54.
(ii) Standard deviation 6·42.

13. ·34

14. ·045

15. ·18

Exercise 9(*d*)

1. (a) $\frac{1}{20}$ (b) $\frac{8}{25}$ (c) $\frac{248}{2475}$ (d) $\frac{1}{1650}$
(e) 0 (Impossible. There is only one in the highest class interval.)
(f) $\frac{82}{495}$ (g) $\frac{1}{1650}+\frac{203}{2475}+\frac{82}{495}+\frac{7}{165}+\frac{1}{495}+0=\frac{161}{550}$

2. (a) $\frac{1}{5}$ (b) $\frac{1}{15}$ (c) $\frac{1}{29}$ (d) $\frac{57}{145}$

3. (a) $\frac{11}{64}$ (b) $\frac{13}{32}$ (c) $\frac{19}{64}$; $\frac{19}{224}$

4. $\frac{11}{336}$; $\frac{5}{372}$

5. (a) $\frac{1}{8}$ (b) $\frac{3}{8}$ (c) $\frac{1}{2}$ (d) $\frac{3}{8}$
(e) $\frac{5}{8}$

6. (a) $\frac{13}{204}$ (b) $\frac{1}{17}$ (c) $\frac{1}{5525}$ (d) $\frac{1}{132600}$
(e) $\frac{8}{16575}$ (f) $\frac{16}{5525}$

7. $\frac{3}{8}$ if he does it the shortest way.

8. (a) $\frac{5}{9}$ (b) $\frac{2}{27}$

9. (a) $\frac{3}{8}$ (b) $\frac{1}{4}$

10. (a) $\frac{2}{5}$ (b) $\frac{2}{5}$

CHAPTER 10

Exercise 10(*a*)

1. *A*: cube, conker, sausage, knitting needle, loaf of bread, glove, electric light bulb, peanut, test-tube, figure seven, copper wire, balloon, Wellington boot, skipping rope.
B: washer, plant pot, bachelor's button, blow pipe, letter R, corn plaster Yale key, book (hole through spine binding).
C: spectacle frames, jacket (ignoring button holes), front door, letter B.
D: all others on the list. (A tea pot without a perforated strainer in the spout would fall into category *C*.)

Exercise 10(*b*)

2. (a) Two sided.
(b) Orientable.
(c) 2 interlocking bands each having two half-twists.
(d) 4 interlocking bands each with two half-twists.

3. (a) One sided.
(b) No.
(c) A continuous band containing a knot.

4. (a) Two sided.
(b) Yes.
(c) Two bands, one of which interlocks twice with the other.

5. (a) One sided.
(b) Two sided.

6. 4 half-twists and orientable; 6 half-twists and orientable; a "four leaf clover fly-over".

Exercise 10(*c*)

1. (a) 6, 8, 12. (b) 5, 6, 9. (c) 7, 10, 15.
(d) 8, 12, 18. (e) 4, 4, 6. (f) 5, 5, 8.
(g) 6, 6. 10. (h) 9, 8, 15.

2. $F + C - E = 2$.

3. $F + C - E = 12 + 12 - 24 = 0$. Torus.

4. $F + C - E = 0$ for the torus.

Exercise 10(*d*)

1. Yes.
4. No.

Exercise 10(*e*)

1. Yes. It is sufficient to remove any one bridge but the starting point will vary.

2. No. Possible from any point if two bridges are removed, one from the left hand group of four bridges and the other from the two on the right hand side.

3. Yes—by adding one bridge from region *P* to region *Q* or *S*.

4. Yes—by adding two bridges, one to the left hand group of four and one to the right hand pair.

Exercise 10(*f*)

1. 2
2. 2
3. 3
4. 4; two ways.
5. 2; one way only.
6. (a) 3 (b) 4
7. Six colours are required as follows:

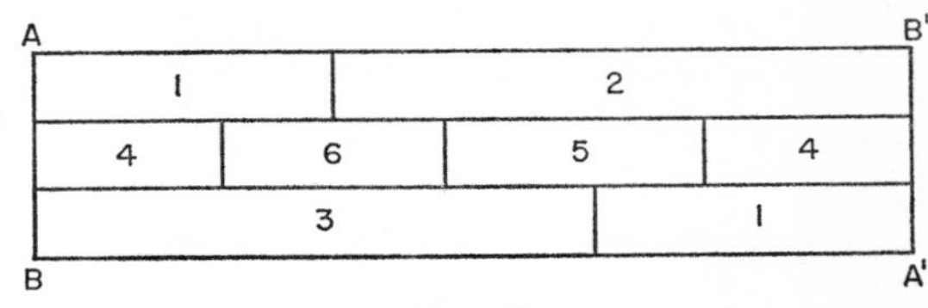

FIG. 10